AF566949

Interfacing Microprocessors in Hydraulic Systems

FLUID POWER AND CONTROL

A Series of Textbooks and Reference Books

Consulting Editor

Z. J. Lansky
Parker Hannifin Corporation
Cleveland, Ohio

Associate Editor

Frank Yeaple
Design News Magazine
Cahners Publishing Company
Boston, Massachusetts

1. Hydraulic Pumps and Motors: Selection and Application for Hydraulic Power Control Systems, *by Raymond P. Lambeck*
2. Designing Pneumatic Control Circuits: Efficient Techniques for Practical Application, *by Bruce E. McCord*
3. Fluid Power Troubleshooting, *by Anton H. Hehn*
4. Hydraulic Valves and Controls: Selection and Application, *by John J. Pippenger*
5. Fluid Power Design Handbook, *by Frank Yeaple*
6. Industrial Pneumatic Control, *by Z. J. Lansky and Lawrence F. Schrader, Jr.*
7. Controlling Electrohydraulic Systems, *by Wayne Anderson*
8. Noise Control of Hydraulic Machinery, *by Stan Skaistis*
9. Interfacing Microprocessors in Hydraulic Systems, *by Alan Kleman*

Other Volumes in Preparation

Fluid Power Design Handbook, Second Edition, Revised and Expanded, *by Frank Yeaple*

Interfacing Microprocessors in Hydraulic Systems

Alan Kleman

Racine Fluid Power, Inc.
Racine, Wisconsin

Marcel Dekker, Inc. New York and Basel

Library of Congress Cataloging-in-Publication Data

Kleman, Alan
Interfacing microprocessors in hydraulic systems
p. cm. – (Fluid power and control : 9)
Includes bibliographies and index.
1. Hydraulic control. 2. Oil hydraulic machinery.
3. Microprocessors. I. Title. II. Series.
TJ843.K54 1989
629.8'042–dc19 89-1614
ISBN 0-8247-8063-9

This book is printed on acid-free paper.

MARCEL DEKKER, INC.
270 Madison Avenue, New York, New York 10016

Current printing (last digit):
10 9 8 7 6 5 4 3 2 1

PRINTED IN THE UNITED STATES OF AMERICA

To my parents

Preface

Few technological developments have been applied as rapidly and widely as the microprocessor. Thirteen years ago, there were no microcomputers or electronic programmable controllers. Today, most of our communication, calculations, and machine control are influenced by these devices.

Although microprocessors can be fascinating, technologists are learning that applying these devices to machine control can be quite difficult and frustrating. The problems experienced by hydraulic engineers and technicians are similar to those faced by their colleagues in other nonelectrical industries. I believe there are two major obstacles to overcome in learning how to interface microprocessors with hydraulic machinery.

The first obstacle concerns the interdisciplinary nature of interfacing. In order to interface successfully, or to control a hydraulic system with a microprocessor-based controller, one must have a working knowledge of the following: microcomputer software, input and output techniques, analog amplifiers, electrohydraulic components, hydraulic systems, and the process being controlled.

The second obstacle involves the vast number of electronic components on the market. It is difficult to find information on the components that one needs without some form of direction or starting point.

This book was written to help hydraulic engineers and technicians solve these problems. Basic explanations are presented concerning the operation of the various electronic and hydraulic components, with an emphasis on the practical knowledge needed to apply these devices. This information will not transform the reader into an expert, but it will enable that person to begin interfacing projects while providing direction for future study.

The structure of this book is modeled after a block diagram of a microprocessor-controlled hydraulic system. Each of the blocks, and the lines connecting the blocks, is a chapter of the book. The first and last chapters present overall system considerations.

Many people have contributed valuable information for this book: Michael Kapp, Peter Bunce, and Allan Key are just three of these people. Many of the illustrations were the work of Roger Kaiser. Finally, Joanne Jarocki provided support and proofreading.

Alan Kleman

Contents

1

Overview

Microprocessor interfacing to hydraulic equipment is a very broad subject. It is interdisciplinary, requiring an understanding of both hydraulic and electronic concepts. To attempt to present, in one book, "how to" methods covering all the techniques used would be impossible. It is also impossible to break down the topic into 10 "rules of thumb" that cover all the methods.

This book attempts to identify and present many of the concepts involved in interfacing microprocessors to hydraulic systems. Discussed in this first chapter are the concepts presented in each of the other chapters.

To organize the varied topics to be presented, a block diagram of a computer-controlled hydraulic system is used (Figure 1.1). The chapters of the book are based upon the blocks or interconnections of the blocks.

The microprocessor is the "brain" of a microcomputer. The other electronic components serve the microprocessor CPU

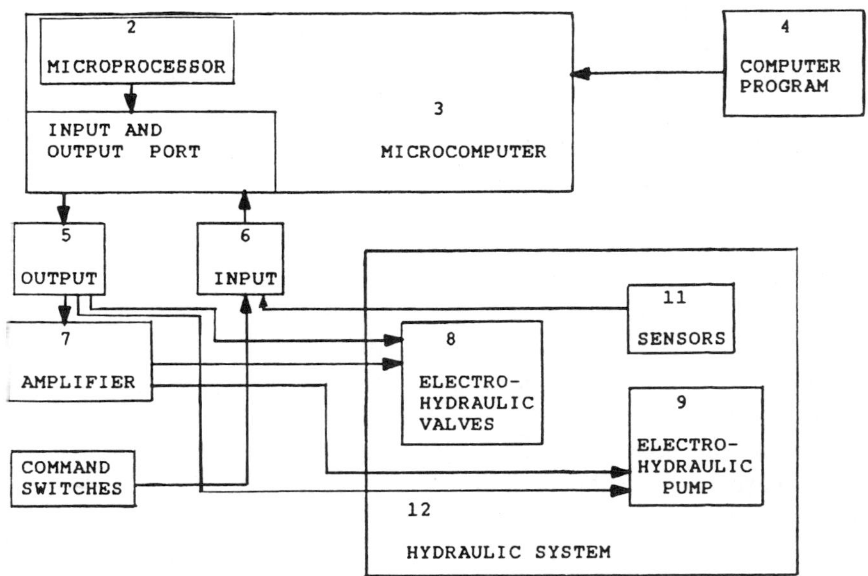

Figure 1.1 Block diagram of a microprocessor-controlled hydraulic system, and outline of this book.

(central processing unit) by storing information and directing the flow of information into and out of it. The microprocessor made possible small and inexpensive computers. In Chapter 2 the operation of a microprocessor is explained, beginning with digital logic circuits. Electronic components, such as transistors, diodes, capacitors, and resistors, can be arranged into circuits that make simple decisions. Binary arithmetic expands the decision-making capability by providing a software system.

The flow of information within and out of the microprocessor is discussed, including the electronic events. Fortunately, an understanding of microprocessor operation is not required to interface these devices in hydraulic systems. Microcomputer manufacturers have done much of the interfacing work for us, providing controllers that require a minimum of microprocessor knowledge.

The importance of Chapter 2 is in demystifying the microprocessor. In taking some of the mystery out of it, microcomputers can be applied with more confidence. Another benefit of this chapter is exposing the reader to a "digital" way of thinking.

Although all microcomputers are similar in their use of a microprocessor, there are great differences in the packaging and I/O (input/output) devices. These differences make some microcomputers more suitable for hydraulic system control than others. In Chapter 3, these differences are discussed, and several microcomputer methods of control are presented. An example of a hydraulic system controlled by a programmable logic controller (PLC) is discussed in some depth.

Many books have been written about all the languages now in use; this book simply directs one's attention to the three "levels" of software. The levels of software are very important to a system designer because of the extensive software development time in virtually all microcomputer-controlled systems. It is true that the software costs usually equal or exceed the hardware costs in a microcomputer-controlled system. Therefore, understanding the differences between the software types may help a designer to avoid a software system that he or she could not implement or service economically.

Chapter 5 is one of the chapters making up the "heart" of this book. The concepts presented are fundamental in controlling a system with a microcomputer. The components discussed begin at the output port of a microcomputer and end with either a discrete solenoid-operated directional valve or an amplifier for an electrohydraulic valve or pump. Also included is a discussion of parallel and serial data conversion techniques necessary in communication over long distances.

Chapter 6 is another key chapter in this book. The methods of conditioning command signals and sensor signals for use by the input port of a microcomputer are presented.

Chapter 7 discusses part of the electronic/hydraulic interface required in all computer-controlled hydraulic systems, the electronic amplifiers that power the solenoids of the electrohydraulic valves. The other part of the electronic/hydraulic interface is obviously the electrohydraulic proportional valves, servovalves, and pumps.

These valve driver amplifiers are usually purchased with the valve or pump to be controlled. The design of these cards is closely tied to the design of the valves, and it is usually cost effective to use these rather than to attempt to design an amplifier.

In Chapter 8 both servovalves and proportional valves are discussed. The difference between these types of valves in price and performance is shrinking. A design example is presented in which a directional four-way valve is redesigned into a proportional valve.

In Chapter 9 concerns the advantages of control of a hydraulic system at the pump over valve control; these are energy efficiency and circuit simplicity. Various methods of electrohydraulic pump control are discussed, along with hydraulic methods. Understanding the methods can aid the designer in understanding the performance specifications of pumps and their capabilities.

Intelligent actuators are primarily subsystems that contain a servovalve, cylinder, and microcomputer in the same physical package. The microcomputer is a dedicated controller for the valve. At present these subsystems represent a small percentage of the hydraulic market, but discussion of these in Chapter 10 is useful in providing a glimpse of what may lie ahead in electrohydraulics.

A discussion of pressure and speed sensors is presented in Chapter 11. Flow sensors are not included because in system design the designer is usually more concerned with actuator speed than flow rate. Actuator speed is also more accurately and economically measured than fluid flow rate.

In Chapter 12 the emphasis is on "what" can be done rather than "how" to do it. Examples of microcomputer-controlled hydraulic systems are presented that show the advantages of this method of control compared with electrical and hydraulic methods. The improvement of an electronic controller (a PID loop controller) by the microprocessor is also discussed.

Caution: Examples of systems are presented for the illustration of concepts. For clarity, it was not practical to show all the interlocks and other safety devices that must be part of any hydraulic system. It is the responsibility of the system designer to be sure that sufficient safety devices are part of the systems that he or she designs and that he or she has done sufficient failure-mode analysis to ensure a safe system.

2

Microprocessors

2.1 INTRODUCTION

If you told a mechanical engineer of the 1950s that you had a machine that contained 20,000 parts, performed 20,000 functions one after another in 1 second, with a reliability of 99.999%, he would not have taken you seriously. This is, however, what a microprocessor does for us.

What is a microprocessor? It is a strip of silicon that contains the equivalent of thousands of transistors (Figure 2.1). The transistors, resistors, and other components are chemically etched into or deposited onto the strip in layers. The strip is typically .25 × .25 inches in size and is mounted in a plastic or ceramic substrate. The substrate can be of several forms. Shown in Figure 2.1 is a 40 pin DIP (dual in-line package) configuration. The substrate contains the conductors between the microprocessor and the input and output pins.

How does it work? The components are connected together in digital logic circuits. Using binary arithmetic, complex coding (software), and two voltage states (generally 5 and 0 Volts), numbers are entered into the microprocessor. Calculations are performed on these numbers by electronic digital logic circuits, and the resulting numbers are sent out from the microprocessor.

Real-life tasks and problems are reduced to thousands of simple calculations by layers of software. The speed of operation of a microprocessor is incredible, performing hundreds of thousands of calculations per second. This speed makes it practical to perform this great number of calculations and translate the results into meaningful information or commands, in a matter of seconds.

This chapter presents an overview of the topics involved in the operation of a microprocessor. It begins with a discussion of the function of digital logic circuits. Binary number systems and arithmetic circuits are then discussed. These topics are the foundation for the discussion of the operation of a microprocessor.

The purpose of this chapter is to acquaint the reader with the concepts involved in microprocessor operation within a microcomputer. Much of the knowledge presented in this chapter is not required in interfacing to hydraulic systems. This is because most

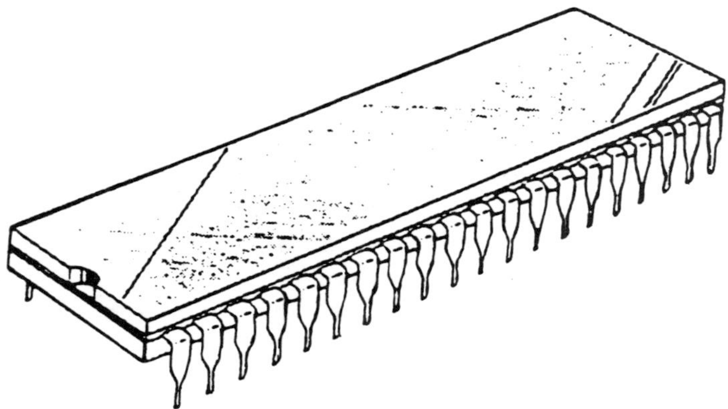

Figure 2.1 Microprocessor.

of this work has already been done for us by the manufacturers of microcomputers. Programmable controllers and industrial computers are available that do a good job of handling all the tasks between the user software and the output port of the controller.

2.2 SIGNIFICANCE OF MICROPROCESSOR DEVELOPMENT

Microprocessors were developed in 1971. This was the breakthrough in miniaturization of electronic components that made possible small and inexpensive computers. Prior to the microprocessor, computers were very large, taking up most of the space in a small room, and expensive. Dedicated computers for machine control or personal computers were not economically feasible until inexpensive microprocessors became available. For historical perspective, a chronology of electronic developments related to the microprocessor is provided in Figure 2.2 [1].

1947	Invention of the transistor
1959	Invention of the integrated curcuit
1971	First 4 bit microprocessor introduced: Intel 4004
1972	First 8 bit microprocessor introduced: Intel 8008
1975	First low-price ($25) 8 bit microprocessor: MOS 6502 (Apple II and Commodore VIC 20 Personal computers)
1979	First 16 bit microprocessor to be widely used is introduced: INTEL 8086 (IBM PC)
1984	Several 32 bit microprocessors introduced

Figure 2.2 Chronology of electronic events related to microprocessors.

2.3 TOPICS INVOLVED IN THE OPERATION OF MICROPROCESSORS

2.3.1 Digital Logic

A microprocessor begins with digital logic circuits. These are electronic circuits that make very simple decisions. In Figure 2.3, a circuit is shown that functions as an AND gate. Both switches A and B must be *on* for the output C (a light) to be *on*.

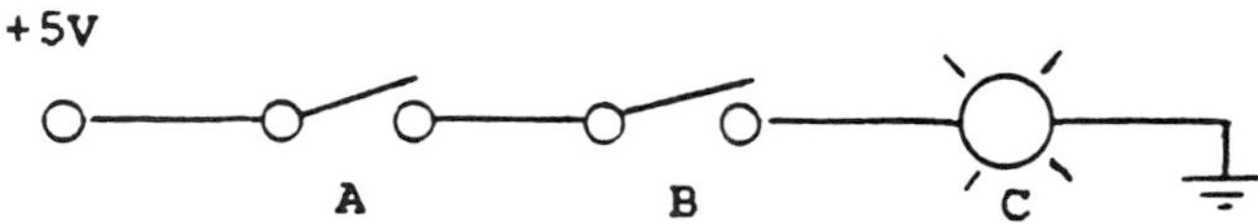

Figure 2.3 AND gate.

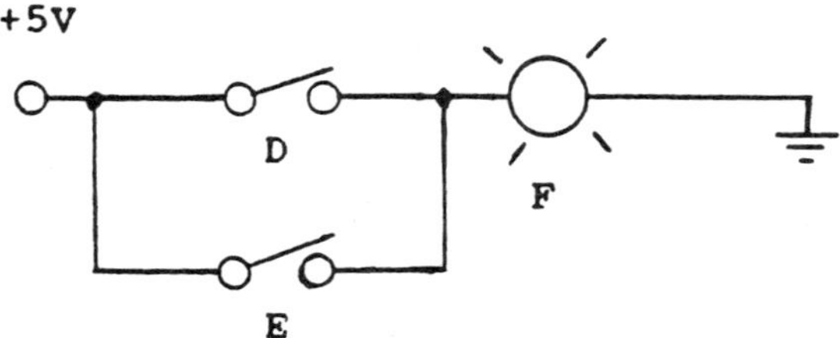

Figure 2.4 OR gate.

In Figure 2.4, another fundamental logic circuit, the OR gate, is represented. If either input switch D or E is on, then the output F will be on.

In Figure 2.5, the third fundamental digital building block is represented, the inverter, or NOT gate. The output H is normally on. If the switch G is on, the current from the 5 Volt source is shorted to ground, stopping the current flow to H. Therefore, activating the input G turns off the output H.

These three gates are the building blocks for all the logic within a microprocessor. Figure 2.6 contains a logic circuit that is a combination of these three gates. Other logic gates are variations or combinations of the first three.

The actual electronic circuits for these gates differ from the simplistic circuits shown in Figures 2.3, 2.4, and 2.5. The actual circuit for a TTL NAND gate can be seen in Figure 2.7.

Transistors are used as switches in this and other gates. A small current, 1 mA for example, passes through the base of the transis-

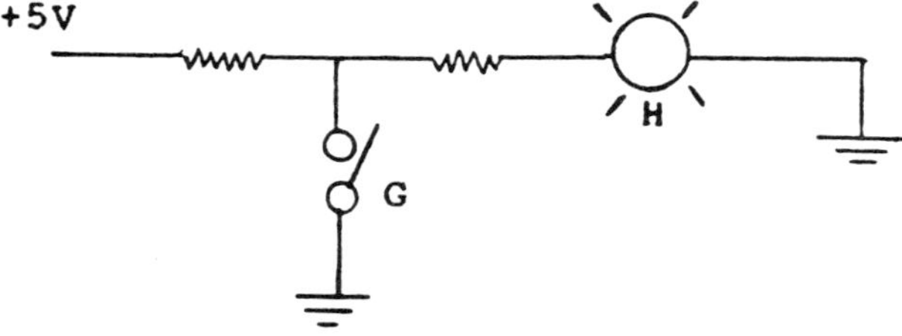

Figure 2.5 NOT gate or inverter.

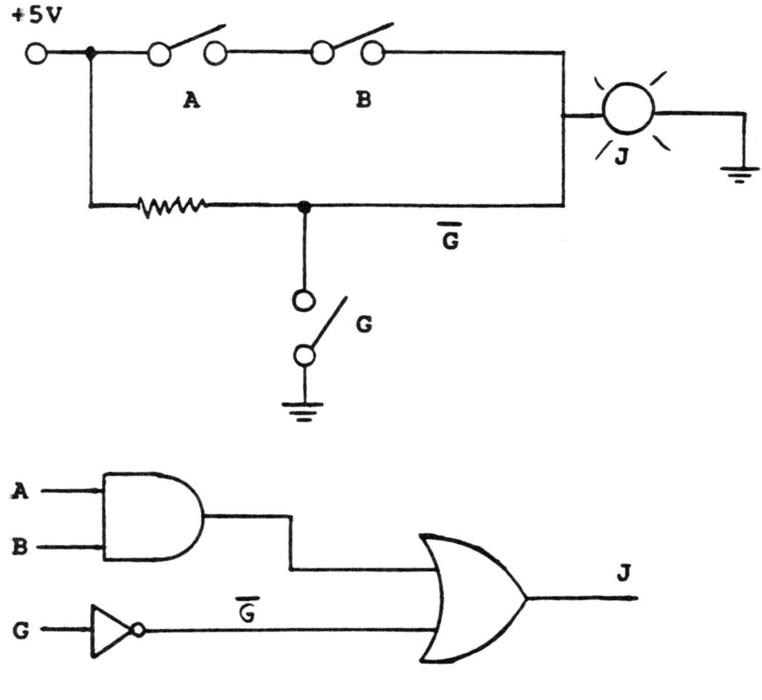

$$J = (A + B) + \overline{G}$$

Figure 2.6 Combination circuit with an AND gate, OR gate, and NOT gate (inverter). J is HI if the quantity A and B is HI or if G is HI.

tor. A larger current, 400 mA for example, from the emitter to the collector is controlled by the flow of the smaller base current.

2.3.2 Combinational Logic

To be useful, these gates must be connected together into complicated logic circuits. To keep track of the logic and to construct circuits that perform desired logic, truth tables and Boolean alge-

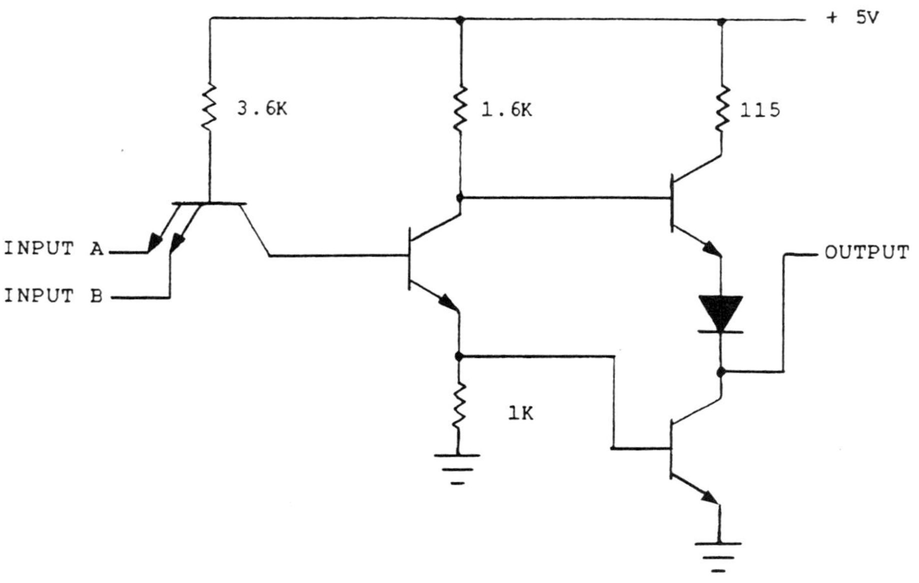

Figure 2.7 TTL NAND gate.

bra are used. The study of these logic gates and the combining of numbers of these gates are referred to as combinational logic. Other topics in combinational logic are Karnaugh mapping and DeMorgan's theorems. These are used to simplify complicated truth tables, thereby simplifying the logic circuits that perform the logic.

Shown in Figures 2.8, 2.9, and 2.10 are the truth tables, Boolean algebra equations, and logic symbols for the three gates. Figures 2.11 and 2.12 show two other important gates, the NAND gate and EXCLUSIVE OR gate. (Note that 1, ON, and HI refer to the same thing, a high-voltage state, and 0, OFF, and LO refer to the opposite, a low-voltage state.)

The NAND gate (NOT AND) is an AND gate with an inverter. This is a useful gate that can be used to construct a great variety of logic circuits.

The EXCLUSIVE OR gate is a useful gate because the middle

INPUT		OUTPUT
A	B	C
0	0	0
0	1	0
1	0	0
1	1	1

TRUTH TABLE

$C = A + B$

BOOLEAN EQUATION

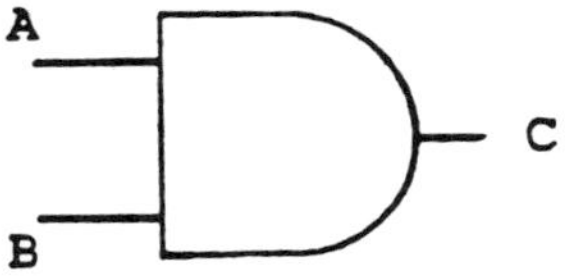

LOGIC SYMBOL

Figure 2.8 AND gate.

INPUT		OUTPUT
A	B	C
0	0	0
0	1	1
1	0	1
1	1	1

TRUTH TABLE

$C = A + B$

BOOLEAN EQUATION

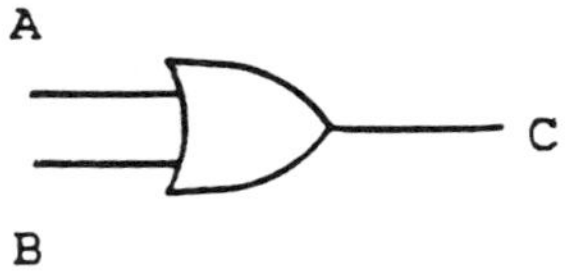

LOGIC SYMBOL

Figure 2.9 OR gate.

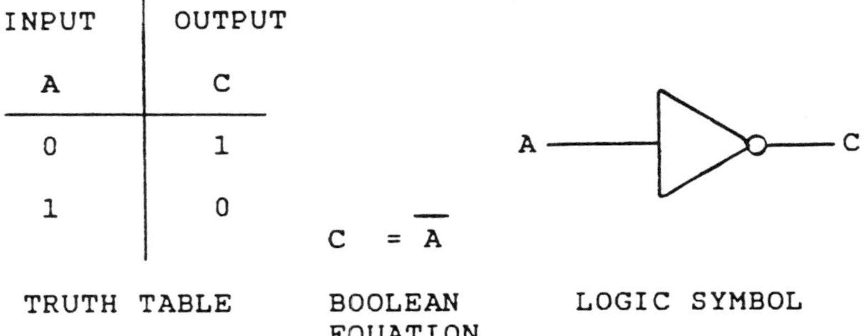

INPUT	OUTPUT
A	C
0	1
1	0

TRUTH TABLE

$C = \overline{A}$

BOOLEAN EQUATION

LOGIC SYMBOL

Figure 2.10 NOT gate.

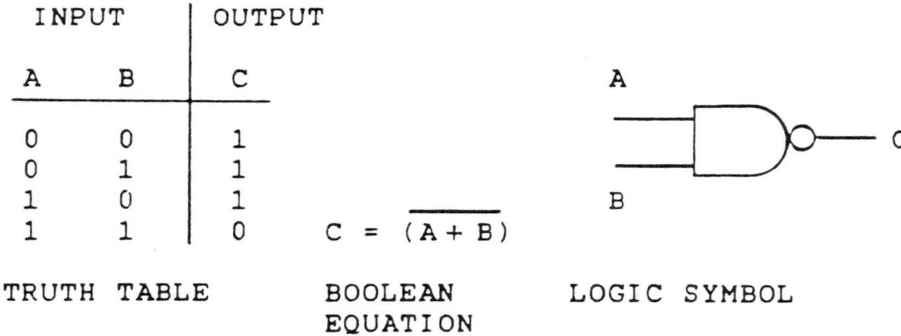

INPUT		OUTPUT
A	B	C
0	0	1
0	1	1
1	0	1
1	1	0

$C = \overline{(A + B)}$

TRUTH TABLE BOOLEAN EQUATION LOGIC SYMBOL

Figure 2.11 NAND gate.

terms are different from the end terms. C is high if either A or B is high. C is not high if both A and B are high.

The main point is that circuits can be constructed from transistors and resistors that have decision-making capabilities. The outputs of these logic circuits or gates change based upon the voltage states (high or low) at the inputs. Numbers of logic gates can be used together to provide quite complicated output, as shown in the following examples.

INPUT		OUTPUT
A	B	C
0	0	0
0	1	1
1	0	1
1	1	0

$C = \overline{(A + B)}$

A
C
B

TRUTH TABLE BOOLEAN EQUATION LOGIC SYMBOL

Figure 2.12 EXCLUSIVE OR gate.

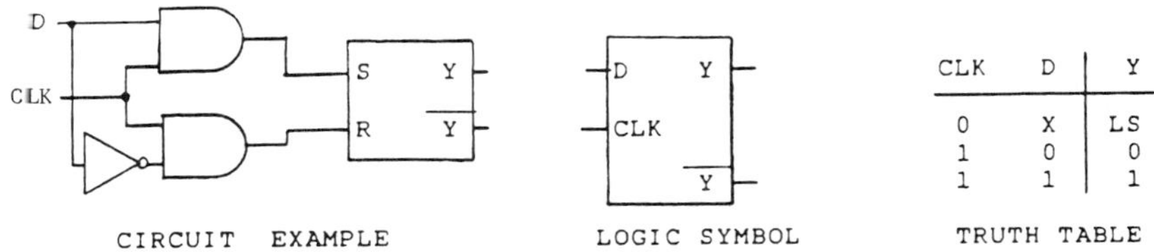

CLK	D	Y
0	X	LS
1	0	0
1	1	1

Figure 2.13 D flip-flop (latch). X = don't care; LS = last state.

2.3.3 Digital Sequential Logic

The flexibility of digital logic circuits was greatly increased with the capability to *store* a result. Circuits could be designed that made decisions based upon past and present inputs. The *flip-flop* is the basic building block used in sequential logic.

A D flip-flop circuit and truth table are shown in Figure 2.13. As long as the clock input CLK is HI, the output changes state every time it receives a pulse at its input D. Therefore, by keeping the CLK input LO, a state can be stored in the flip-flop for an extended period of time. Figure 2.14 shows a circuit that stores a 4 bit number using four D flip-flops. A 7475 is an IC (integrated circuit) chip that contains four flip-flops.

Counters are an important functions achieved with flip-flops. In Figure 2.15 a "ripple" binary counter is shown. It is constructed

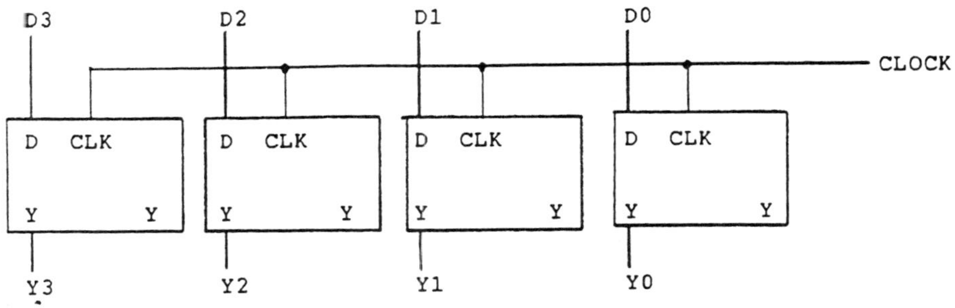

Figure 2.14 Storing a 4-bit number with D flip-flops.

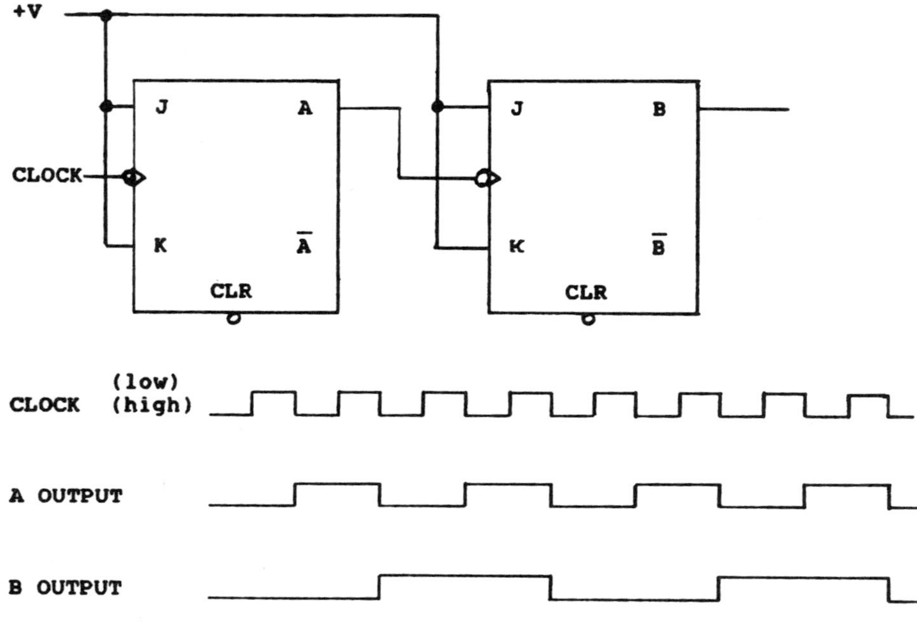

Figure 2.15 Ripple counter circuit and waveforms.

from J-K flip-flops. The state of A must change twice for the state of B to change once. Therefore each flip-flop divides the count by 2. (See Figure 6.10 for an application of a counter.)

2.3.4 Arithmetic Circuits

Digital logic circuits can solve simple logic problems. However, to enter a complicated problem, solve it, and output the solution that can be decoded to something useful, a software system is needed. This system is a binary, base 2 number system. Using a base 2 number system, it is quite convenient to add or subtract numbers with digital logic circuits. Counting in several number systems is shown in Table 2.1. This is a concept fundamental to the way electronic digital circuits perform calculations. The binary

Table 2.1 Comparison of Three Number Systems

Decimal (base 10)	Binary (base 2)	Hexadecimal (base 16)
0	0000	0
1	0001	1
2	0010	2
3	0011	3
4	0100	4
5	0101	5
6	0110	6
7	0111	7
8	1000	8
9	1001	9
10	1010	A
11	1011	B
12	1100	C
13	1101	D
14	1110	E
15	1111	F
16	10000	10

arithmetic shown is the basis for all microprocessor calculations and operations. It transforms the voltage-level outputs of transistor circuits to a numerical language that can be used for solving problems.

The addition of two binary numbers is accomplished through adder circuits. The logic circuit and truth table for a half-adder are shown in Figures 2.16 and 2.17. The circuit adds two binary numbers and outputs the sum and the carry digits.

The half-adder is limited in that it can only add two digits at a time. Full adders are more commonly used and are made up of several half-adders. Shown in Figures 2.18 and 2.19 are the logic circuit and truth table for a full adder. Note that the carry and sum digits are a binary representation of the decimal number. In other words, in the last row, the 1 and 1 are a 2 bit binary representation of the number 3. Full adders are used in combinations to produce 4 bit and larger adders.

INPUT		OUTPUT	
A	B	Carry	Sum
0	0	0	0
0	1	0	1
1	0	0	1
1	1	1	0

Figure 2.16 Truth table for a half-adder.

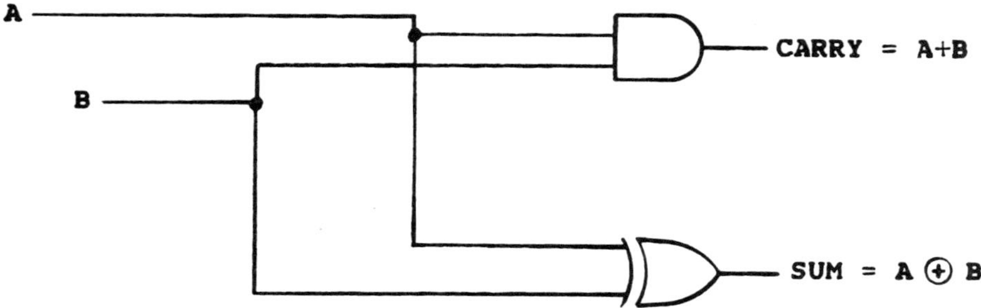

Figure 2.17 Half-adder circuit.

INPUT			OUTPUT		EQUIVALENT NUMBER
A	B	C	CARRY	SUM	(BASE 10)
0	0	0	0	0	0
0	0	1	0	1	1
0	1	0	0	1	1
0	1	1	1	0	2
1	0	0	0	1	1
1	0	1	1	0	2
1	1	0	1	0	2
1	1	1	1	1	3

Figure 2.18 Full adder truth table.

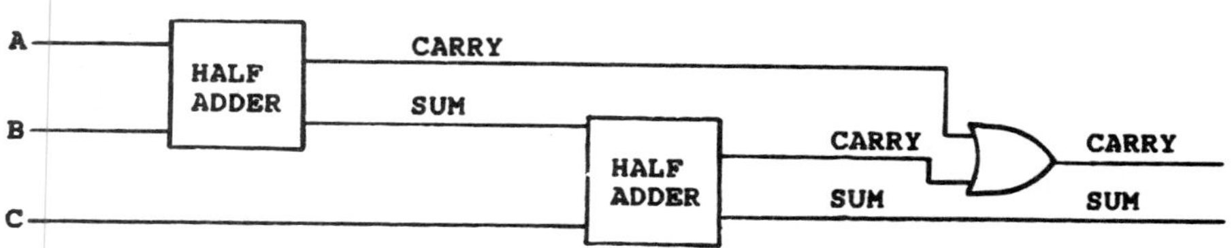

Figure 2.19 Full adder circuit.

A discussion of binary arithmetic and digital logic shows us how electronic circuits can be used to perform calculations and execute a sequence of events. Discrete digital logic circuits were used for these purposes prior to the development of the microprocessor.

Combinational and digital logic circuits can be used to make decisions and store a result. However, to perform a complicated calculation, large numbers of digital components connected in complex circuits are required. The number of circuits and the complexity of the logic circuits become unmanageable even for simple problems. A system was needed that could rapidly "construct" a digital logic circuit, solve a problem, and dismantle the circuit and construct another. Software does this for us.

2.3.5 Digital Logic Using Software

To understand how a software system replaces hard-wired logic, we must begin at the basis for all calculations and functions of a microprocessor, the binary number system. This is the "language" the microprocessor uses. All instructions, data, and addresses (locations) are in the form of two voltage levels that represent the numbers 1 and 0 of a binary (base 2) number system.

2.3.6 The Instruction Set of a Microprocessor

Binary arithmetic is the language, but a system is needed. This is provided by the instruction set of the microprocessor. The instruction set of a microprocessor is a set of instructions in the form of 8 bit (in an 8 bit microprocessor) binary numbers that command the microprocessor to perform all the operations of which it is capable. These operations are:

Fetching numbers from memory location
Storing numbers in temporary memory locations (registers)
Performing arithmetic in registers and sending the resulting numbers to memory locations or to output ports
Responding to hardware interrupts

Statements from the instruction set are arranged in a sequence referred to as a program. Shown below is a statement from the instruction set of a common microprocessor, the Intel 8085A.

Statement	Comments
LDA	Mnemonic form of the statement
0 0 1 1 1 0 1 0	OP CODE
1 0 1 0 1 0 1 0	LOWER-ORDER ADDRESS
1 1 1 1 0 0 0 0	HIGHER-ORDER ADDRESS
Cycles: 4	

Mnemonics are a code used for the convenience of programmers. These statements are abbreviations of statements from the instruction set. Mnemonics are used in programming in assembly language and are converted into machine code (1s and 0s) by an assembler program.

LDA is a mnemonic for the statement "load accumulator direct." The content of the memory location specified in the second and third bytes is moved to register A.

The OP CODE contains 8 bits and is the first byte of the instruction. This byte contains the instruction, defined in the instruction set, that the microprocessor will execute. This may be the only byte in the instruction or there may be one or two more containing data or an address.

The lower and higher order bytes are combined to form one 16 bit address. In this example the lower order byte (1010101010) and the higher order byte (11110000) would be combined into the 16 bit address (1111000010101010). By using 2 bytes for an address, the 8085 microprocessor can address 65,536 memory locations (64K).

There are 92 other instructions in the 8085 instruction set. The statements perform the functions listed here:

Data transfer statements transfer data between registers and between memory locations and registers, exchange data between registers, and provide temporary storage of data.

Arithmetic statements perform addition, subtraction, and incrementing and decrementing data in registers or memory.

Logical operations are performed on data in registers or memory. These operations include AND, OR, XOR, comparisons, and complements.

Branch statements cause the execution of the program to jump from one section of the program to another and to return.

Stack, I/O and machine control statements control the stack (six registers in the microprocessor), data transfer to and from output and input ports, and other control functions.

2.3.7 Speed of a Microprocessor

The LDA statement requires four machine cycles to execute. An 8085 AH-1 microprocessor has an instruction cycle time of .67 ms. Therefore, the LDA statement requires

$$4 \times .67 \text{ ms} = 2.68 \text{ ms} = .00000268 \text{ s}$$

to execute.

This calculation can give some insight into the speed at which a microprocessor can process data to control a hydraulic system. However, to estimate the speed of response of an entire system (a microcomputer, electrohydraulic amplifiers, and hydraulic actuators), many other factors must be taken into consideration: other functions being performed by the microprocessor, the programming language used, the time for data transfer to external devices; D/A (digital-to-analog) and A/D conversions, the response time of

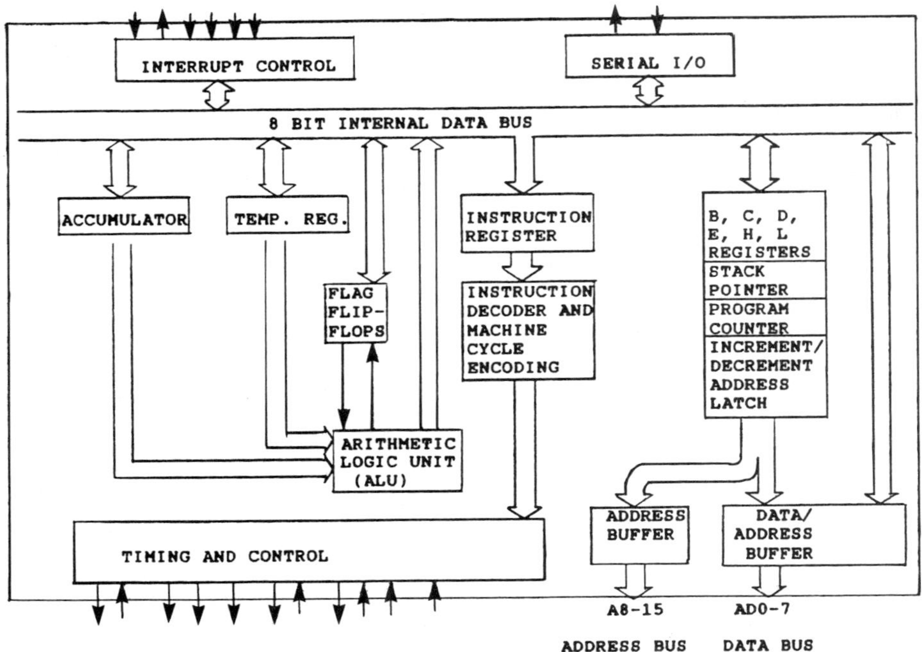

Figure 2.20 CPU of a microcomputer: block diagram of an INTEL 8085A microprocessor.

the hydraulic actuators, and the volume of fluid in the controlled sections of the circuit.

2.4 INSIDE A MICROPROCESSOR

A typical microprocessor contains the components shown in Figure 2.20. This is a popular 8 bit microprocessor, the Intel 8085A. Following are comments regarding the components of the 8085A. Other microprocessors may differ in the arrangement of the components, and specific product literature should be consulted if detailed information is needed.

The central processing unit (CPU) in a microcomputer performs all the logic and control functions. The CPU is typically a single microprocessor and contains the following functional units:

An arithmetic logic unit (ALU)
Registers for the temporary storage of numbers
Control circuitry
Internal bus

2.4.1 Arithmetic Logic Unit

In this section of the microprocessor, calculations take place. Addition and subtraction are performed, along with extensions of these, division and multiplication. Other logical operations are also performed, such as masking, gating, and manipulations of the contents of registers.

2.4.2 Registers

Registers are temporary memory locations within the CPU that are used to store data, addresses, and timing information.

The Accumulator

The accumulator is a general-purpose register that contains one of the numbers involved in a calculation being done by the ALU. In a typical program step, a number may be fetched from memory and added to the number in the accumulator. The accumulator then stores the resulting number. In general an accumulator is used as both a source register and a result register.

Additional general-purpose registers often perform the function of an accumulator. Data transfer among these registers is much faster than transfer between the CPU and external memory chips. Therefore, these registers can be used to increase the speed of the microprocessor.

The Program Counter

During the execution of a program, it is necessary for the CPU to keep track of which program step is being executed and be prepared to fetch from memory the next program step. The program counter is a register that contains the address (memory location) of the next program step to be executed. The program step being executed is stored in the instruction register.

2.4.3 Control Circuitry

The control circuitry of a CPU performs the functions of timing of the program execution and the direction of data flow within the microprocessor and between the microprocessor and external support chips. The control circuitry directs the flow of data by switching the appropriate chips to a write, read, or inactive mode.

2.4.4 Internal Bus

The parts of a microprocessor are interconnected by a group of eight conductors, an internal BUS. All data transfers within a microprocessor are done on the same eight conductors (in an 8 bit microprocessor). Considering the speed of operation of a microprocessor and all the data transfers occurring on the same conductors, the importance of timing is obvious.

2.5 OUTSIDE THE MICROPROCESSOR

For a microprocessor to be useful it must be connected to additional external digital logic circuits. It must have memory chips in which to store data and input and output chips to hold the data and transfer it to the output ports of the microcomputer. A microprocessor in a microcomputer is shown in the block diagram of Figure 2.21.

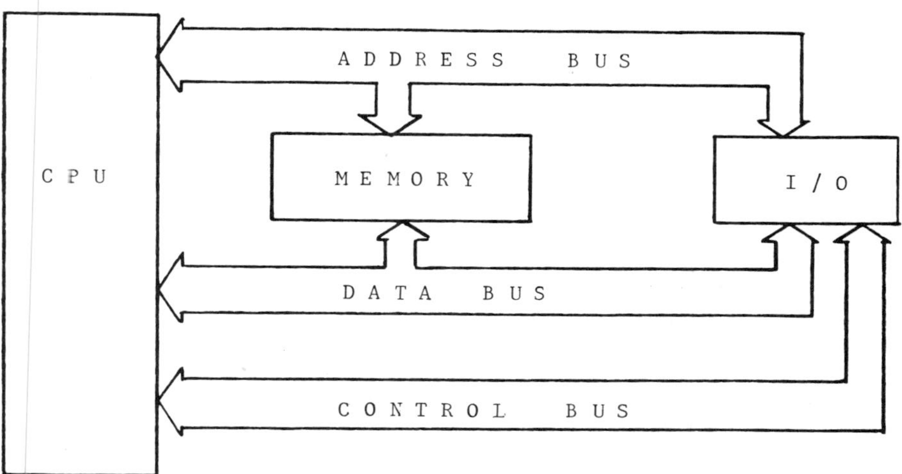

Figure 2.21 Microcomputer block diagram.

2.5.1 The Clock

This is an electronic oscillator that generates a series of voltage pulses. It can be part of the CPU or an external chip. It sets the timing for all the operations within a microcomputer. The frequency of the pulses can be quite high, 12,000,000 cycles per second in some microcomputers (12 MHz).

2.5.2 Buses

A microprocessor is connected to other chips by three groups of conductors (Figure 2.21), called buses. In the 8085A there is the 8-conductor DATA bus, the 16-conductor ADDRESS bus, and the 4-conductor CONTROL bus. All the external chips and the CPU are connected to the buses. The data bus carries data between the CPU and the memory and I/O (input/output) chips. The address bus carries information used to direct the data flow to the correct memory location on the correct chip. Address information is decoded to activate the correct chip and to electrically connect the correct memory location to the I/O pins of the chip. The control bus activates the READ or WRITE line of the specific chip that has been chosen to exchange data with the CPU. Only one chip can be in a READ state and only one chip can be in a WRITE state at one time. The other chips are in an OFF condition.

2.5.3 Memory

Memory chips are digital integrated circuit chips used to store data or programs. These chips are connected to the CPU by buses (data, address, and control buses). There are several types of memory chips classified according to programming and storage capability.

ROM (read-only memory) is used for permanent or semi-permanent storage of programs or data. A ROM chip retains its data indefinitely without power being applied to it. The microprocessor can receive data from a ROM chip but it cannot send data to it. Some ROM chips are programmed during the manufacturing process using masking techniques, and others (PROMs and EPROMs) are programmed afterward.

PROM (programmable read-only memory) can be programmed, but special equipment is required. The programming requires a high voltage (12–25 V) and takes a considerable amount of time. A PROM holds its data for years without power being applied to it.

An EPROM is an erasable PROM. The erasure is accomplished by shining an ultraviolet light through a small window on the chip. EPROMs are often used to store programs for programmable controllers as an alternative to a disk drive. EPROMs also retain their data indefinitely without power. This capability is referred to as "nonvolatile" memory.

RAM (random-access memory) can be programmed by a microprocessor, and the data can be read by the microprocessor. The programming, referred to as "writing to memory" is rapid, and the microprocessor uses RAM for most of its temporary storage tasks. RAM is volatile, in that it will lose its data if power to the RAM chip is interrupted.

2.5.4 Input and Output Ports

Most external devices that interface with a microprocessor cannot do so directly at the buses of the microprocessor. There are several reasons for this.

One reason is that the external device may be operating at a slower speed than the microprocessor and it is not quick enough to respond to the signal from the microprocessor. Therefore, it is necessary to "hold" a binary number at a value for a period of time so that the external device can carry out its activity. If there no chip held the number the microprocessor would continually send the same number to the bus, preventing it from performing other tasks or executing the program.

Other reasons are the electrical limitations of the microprocessor. The microprocessor may not have the current capability to drive the external device. The output chip acts as a buffer between the outside world and the microprocessor. The output chip can drive a larger load than the microprocessor.

A similar condition occurs with input data. As data enter the microcomputer from the outside world, they must be held until the microprocessor is ready to use the data.

The input and output chips that connect the CPU to the outside world are parallel interface adaptors (PIAs) and universal asynchronous receiver transmitters (UARTs). The PIA and UART are connected directly to the I/O ports of the microcomputer.

2.5.5 Parallel Interface Chips

A parallel interface adaptor (PIA) is a chip found in most microcomputers. It handles the transmission and receiving of parallel data between the CPU and the I/O port. It has another function of holding (latching) a byte of data before input into the CPU or holding a byte after it has been output from the CPU.

2.5.6 Serial Interface Chips

Data within the microprocessor are in parallel form. The transmission of data between microcomputers and external devices is often done in serial form (1 bit at a time). Therefore, incoming serial data must be converted from serial to parallel, and outgoing data must be converted from parallel to serial. Also, the data must be held (latched) at times to wait for the CPU or external device to accept it. A UART is the chip that performs these tasks. Serial and parallel data conversion are discussed in Chapters 5 and 6.

2.6 MICROPROCESSOR OPERATION

All the tasks a microprocessor performs begin with two actions. The microprocessor *fetches* an instruction from a memory location and executes the instruction. The execution of the instruction may simply be moving a number from one location to another or an arithmetic task, such as addition or subtraction.

Electrically, what occurs during a fetch? The 8 bits in a memory location are connected to the 8 bit locations of a register in the microprocessor. The 8 bits of the register are forced HI or LOW according to the states of the bits in the memory location. A small amount of electric current flows from the memory location to the register, bringing the register to the same voltage state (HI or LO) as the memory location.

In a READ operation, the CPU reads the contents of a memory location. To accomplish this the CPU signals the appropriate memory chip via the control bus to switch to a write (transmit) state. It also sends an address on the address bus, which the memory chip receives. The memory chip then connects the addressed memory location to the data bus, changes the logic states of its bits to the data byte on the data bus.

In a WRITE operation the opposite occurs. The logic states of a register in the CPU are transferred to a memory location. The memory chip is signaled to receive by the CPU. The memory chip receives an address on the address bus from the CPU.

The CPU, memory chips, and I/O chips are connected to the buses at all times. For information to be sent by the correct chip and received by the correct chip, the other chips must be signaled to "ignore" the data transfer. This is done by the off state of the chips. The off state is a high impedance connected to the bus through which current and data cannot flow. In summary, there are three states the chips and the CPU use to control data flow: read, write, and off.

2.7 OTHER MICROPROCESSORS

Other microprocessors should be mentioned. The Z80 by Zilog is one of the most widely used microprocessors. It is an 8 bit chip and was the dominant chip used in personal computers prior to the IBM PC. Now it finds use in dedicated control applications. The Motorola 68000 is a 16 bit chip that is a competitor to the Intel 8086. It is used in the Apple Macintosh, Atari 520 ST, and Commodore Amiga personal computers. It is also used in high-performance workstations, Unix-based multiuser applications, and dedicated control applications.

2.8 MICROCONTROLLERS

Developing in parallel with microprocessors were single-chip microcomputers. These devices contained a microprocessor and the support chips required by the CPU. ROM, RAM, and I/O ports are on the single chip.

The 4 bit microcontrollers have been in use since the mid-1970s. The first of these was the Texas Instruments TMS 1000. It had 1 K of ROM, sixty-four 4 bit words in RAM, a 4 bit input port, and 19 bits of output.

The 4 bit microcontrollers were very crude, with a small instruction set. Programming was difficult and had to be done by masking during the manufacturing process. These devices were cost effective in such cost-sensitive applications as toys and appliances. The 4 bit microcontrollers have been replaced by 8 bit microcontrollers as the cost of these dropped.

The 8 bit microcontrollers have expanded instruction sets and external buses for communication with external ROM or RAM. Many 8 bit microcontrollers are supplied without ROM, with an IC socket on top of the microcontroller chip. This allows the flexibility of producing several controllers with the same microcontroller and different ROM chips or EPROMs.

Also available in 8 bit microcontrollers is the capability to use higher level languages. For example, the Intel 8052 is available with a BASIC interpreter in the on-chip ROM. Motorola's 6801 series and the Zilog Z8 are two other 8 bit microcontrollers.

The 16 bit microcontrollers are complex devices with limited usage compared with the 8 and 4 bit chips. There is one highly visible use of a 16 bit microcontroller—controlling an automobile. Ford Motor Company's fourth generation engine controller, the EEC-IV, uses an Intel 8096 microcontroller.

2.9 SUMMARY

The topics of this chapter were presented to familiarize the reader with microprocessor operation within a microcomputer. For those of us interfacing microcomputers with hydraulic systems, the topics of this chapter are educational but not essential. Because a wide variety of microcomputers on the market can be used to control a hydraulic system, designing and assembling our own microcomputer is not necessary.

Those interested in controlling with a microprocessor and the minimum number of support ICs will find a long road ahead of them. The knowledge required is extensive. A place to begin is with a microprocessor training kit, such as that sold by Heathkit. Microprocessor evaluation boards, containing a microcomputer on a board with a keyboard and prototyping breadboard, are sold by microprocessor manufacturers. Courses in assembly code programming are also a good place to start.

REFERENCE

1. Garetz, Mark, Evolution of the microprocessor, *BYTE Magazine*, McGraw-Hill, New York, September 1985.

BIBLIOGRAPHY

Malvino, Albert Paul, and Leach, Donald. *Digital Principles and Applications.* McGraw-Hill, New York, 1981.

Slater, Michael. *Microprocessor Based Design,* Mayfield Publishing, Mountain View, California, 1987.

3

Microcomputers

3.1 INTRODUCTION

All microcomputers have one thing in common, a microprocessor CPU (central processing unit). Microcomputers differ greatly in the hardware that serves the microprocessor, the software, and the peripheral devices. The differences are so great that microcomputers are referred to by different names depending upon the hardware configurations. Programmable logic controllers, board-level computers, and personal computers are all microcomputers. All of these devices can also be used to control a hydraulic system.

All these microcomputers are discussed in this chapter. Emphasis is on the most common microcomputer used to control hydraulic systems, the programmable logic controller (PLC).

3.2 PERSONAL COMPUTERS

Personal computers (PCs) are attractive as controllers because of the great calculating power that one receives for a relatively small price (Figure 3.1). However, the system designer quickly becomes aware of the shortcomings of PCs used to control machinery.

Consider a PC with a serial output port and the task of activating the solenoid of the four-way valve shown in Figure 3.2. In the PC program, we can output a number to the port, which is then transformed into a series of bits containing the "data."

The first problem is the polarity and voltage of the serial signal. The voltage is too low for a 115V ac solenoid. If it were a 12V dc solenoid, the -15V signal could not be used.

The second problem is that the serial data stream has little current capacity. Even if the serial stream had enough pulses in it to maintain an average voltage approaching the solenoid requirement, the current available from a serial port is less than 15mA. This is far less than the 9 Amp of inrush current that can be drawn by an industrial solenoid valve.

This problem could be overcome by amplifying the signal with a power transistor or by activating a relay, but there is another problem: at what binary number does the valve activate? A crude

Figure 3.1 Personal computer.

relay circuit may discern between a serial signal of nearly all 0s and all 1s, but what about the signals between these?

Another problem when using a serial data stream as a stream of voltage pulses is how does the "valve" in the example tell the difference between two serial numbers 11010000 and 00000111 (Figure 3.3)? The average current that results from this serial data stream is similar, but the binary values are vastly different! We begin to see the serial output as a means for transmitting information rather than transmitting electrical energy.

The final problem is how can we control both solenoids of the valve in Figure 3.2 from the same port? It is impractical to add another output port to a PC for every output device in a system.

Controlling the valve in the example was a dismal failure with a serial port. What about a parallel port? A simplistic method can be used with a parallel port, which controls a number of valves equal to the number of bits in the port (i.e., an 8 bit parallel port would control 8 valves).

Each pin in the port controls a driver (power transistor or relay) that switches current to a solenoid valve. An example is

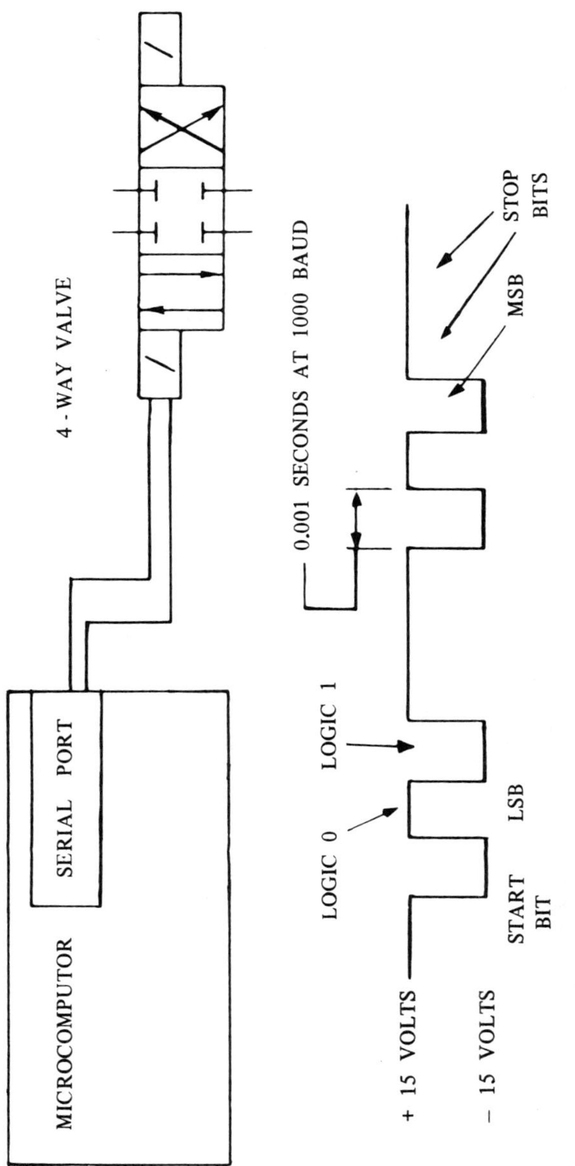

Figure 3.2 Difficulty of controlling with a serial data stream.

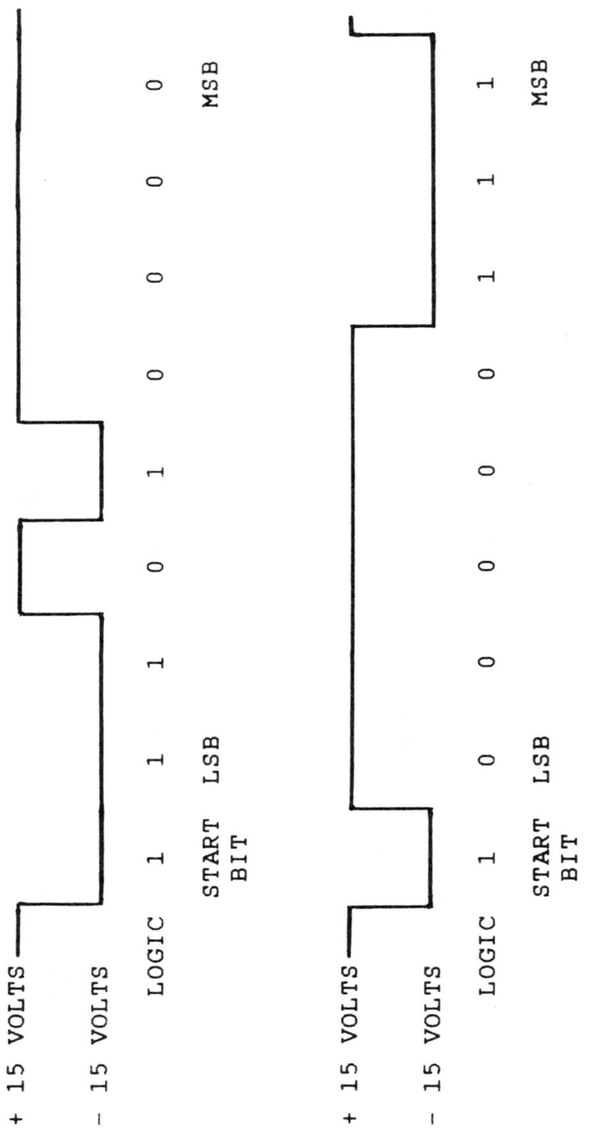

Figure 3.3 Serial data comparison.

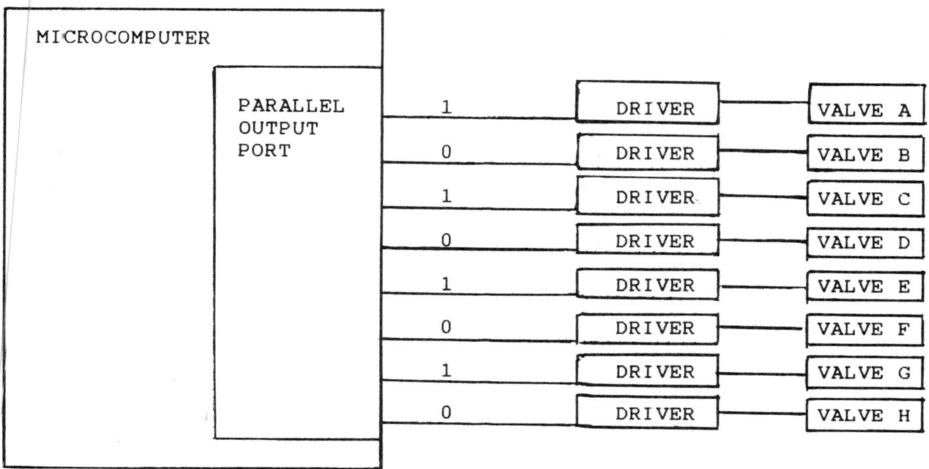

Figure 3.4 Parallel port controlling a valve with each output pin.

shown in Figure 3.4, in which the number 10101010 activates the valves A, C, E, and G. If the number at the port were changed to 10100010, valve E is deactivated, leaving A, C, and G on. This is an improvement over the serial port, but the number of outputs driven by one parallel port is still low, one port for eight outputs.

Controlling a large number of output devices with a single output port is done with addressing. Address bytes and data bytes are sent alternately through the same cable to output devices. (Figure 3.5). The output device receives the first byte of serial data and determines which of the valves is being addressed based upon the value of the binary number contained in the first byte. The second byte is then received by the output device, and one of several events occurs.

In the simplistic example of a single valve, the valve is activated if the second byte is the binary number 00000001. If the binary number of the second byte is 00000000, then the valve is deactivated. (It could be programmed to activate the valve on any numeric value.)

In the case of a proportional valve or servovalve, the binary number of the second byte could contain a pressure or flow rate

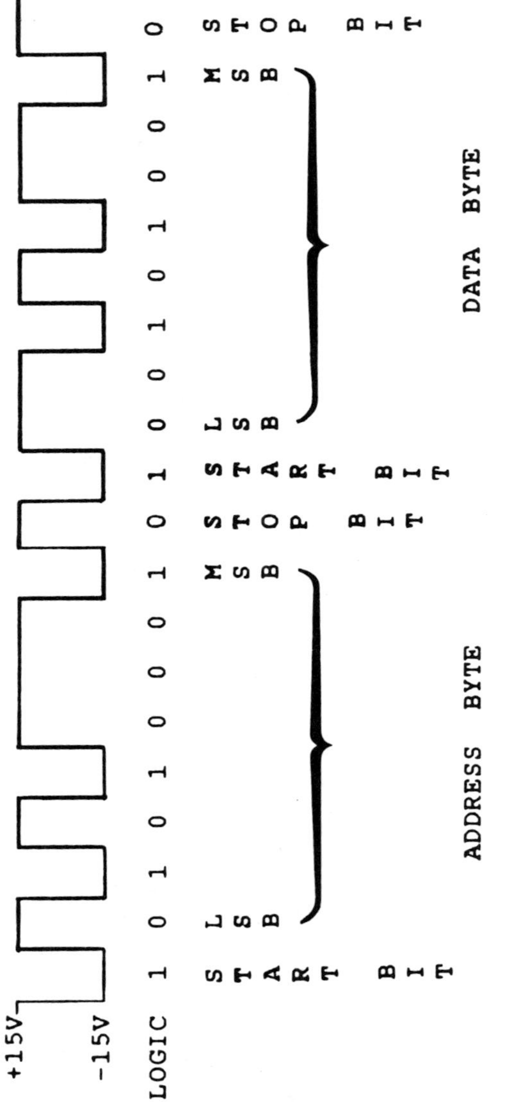

Figure 3.5 Serial output containing address and data information.

set point. (It would have to be latched at this value, converted to an analog voltage. The voltage would command the valve amplifier card.)

Using an addressing method, a large number of devices can be controlled by a single output port. It is possible to control as many as 255 valves with a single 8 bit serial port and 2047 valves with a 12 bit port.

In summary, a personal computer alone cannot control a hydraulic system. It has more than enough calculating power but it lacks the tools—the input and output devices.

Another shortcoming of PCs is the lack of software for machine control. The common PC languages are too difficult to use for many machine designers. Too much programming knowledge is required. Especially difficult is the communication software between the PC and the external cards.

3.3 PC INTERFACES

Some manufacturers of interfaces for PCs have solutions for both the I/O and the software shortcomings of PCs. The Metrabyte Corporation manufactures a line of interface cards for IBM PCs. Included with the cards is software that eases the programmer's task in machine control. In the following examples output cards from the Metrabyte Corporation are used to control a directional valve and a proportional pump.

Figure 3.6 is a diagram of a Metrabyte industrial control system. It utilizes an expansion card (a Metrabyyte MDB-64 driver board) mounted in an expansion slot within an IBM PC. It communicates directly with the PC address, data, and control buses and can control up to 512 output devices with a 50 pin cable.

The IBM PC in the example is programmed in BASIC. The MDB-64 driver board responds to four addresses in the computer's input and output expansion address space. The program writes and reads data from the MDB with BASIC INP and OUT statements. A BASIC program to activate two valve solenoids is shown in Figure 3.7.

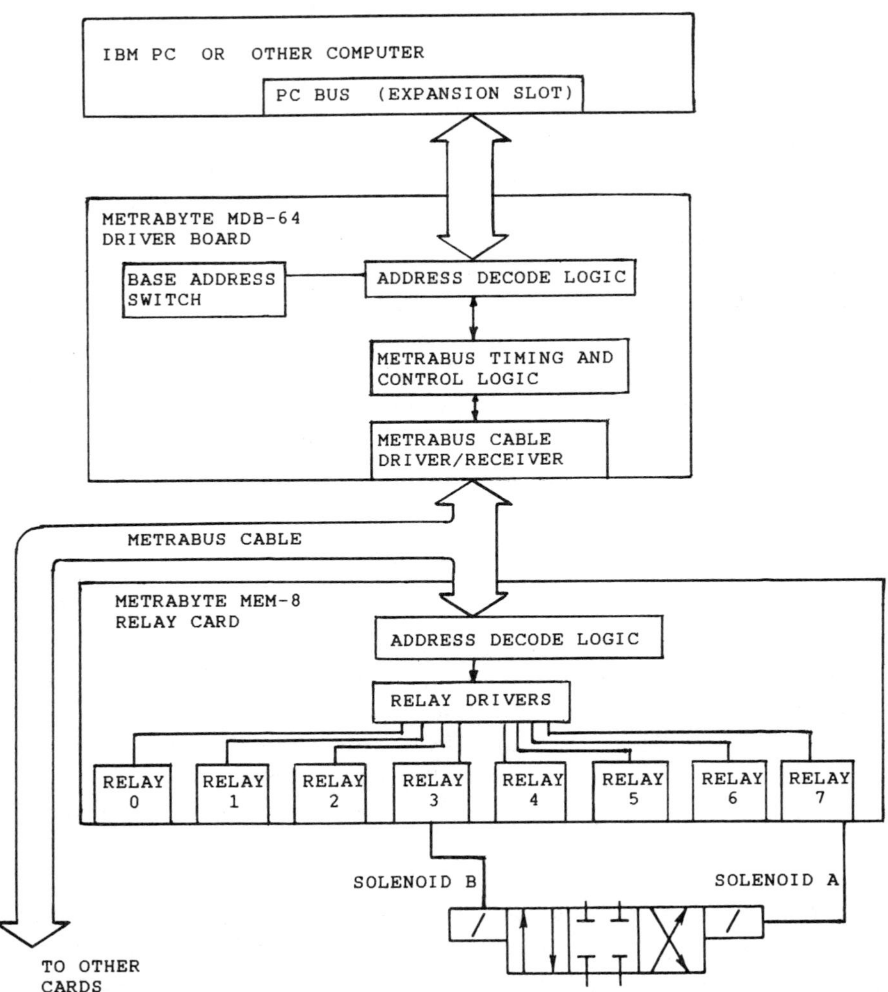

Figure 3.6 Block diagram of Metrabus control system controlling two solenoids.

The Driver board and MEM 8 board are assigned addresses in statements 10 and 20.

10 BASADR = 768	Driver board is set to an address unused by the PC
20 BUSADR = 20	Sets the address of the MEM 8

Statements 30–100 activate solenoid A, deactivate solenoid A, and activate solenoid B.

30 OUT (BASADR + 1), BUSADR	Selects address of MEM 8 board
40 OUT (BASADR), 128	Activates relay to solenoid A
50 FOR L = 1 to 5000	Remain ON for 5
60 NEXT L	
70 OUT (BASADR), 0	Deactivate solenoid A
80 FOR L = 1 to 1000	Pause 1
90 NEXT L	
100 OUT (BASADR), 8	Activate solenoid B

Figure 3.7 Program to activate two discrete output, directional valve solenoids.

The BASE ADDRESS SWITCH in the MDB 64 is a DIP switch that sets the address of the MDB-64 to an unused address of the PC. In this example it is 768. On the MEM-8 relay card the dip switch has been set to the binary number equivalent to "20." This is an address that the MDB-64 uses to activate a relay on the proper relay card.

The relay is selected in statement 40. The binary number 10000000 (128) turns off all the relays except the one controlled by the most significant bit, relay 7. The binary number 1111011 (239) has the same result provided no other valves are attached to the other relays. The binary number 100010000 results in a serious problem: both the solenoids are energized simultaneously.

In statement 100, solenoid B is activated by the binary number equivalent to the decimal number "8." That number is 00001000.

The Metrabyte MAO-8 analog output board can be used to control the output flow rate of the proportional pump in Figure 3.8 by sending a 4–20 mA signal to the pump's amplifier card. The program of Figure 3.9 commands the pump to 100% flow and then to 50% flow.

A discrete input module is also shown in Figure 3.10 to accept the contact closure input from the cylinder LIMIT switch. This is the Metra MII-32 logic level input board. The program of figure 3.11 continually polls the limit switch, testing for contact closure, and causes the program to jump to statement 100 as contact is made.

3.4 BOARD-LEVEL MICROCOMPUTERS

In the mid-1970s, manufacturers of microcomputers began to market "component" microcomputer systems. These systems consisted of a board with a microprocessor CPU and a series of circuit boards with different tasks, such as communication with external devices, graphics for monitors, analog input and output, and floppy and fixed disk controllers.

The important point in these systems was a standard bus to which all the boards were connected. The electrical hardware and the communication software were standardized, permitting designers to connect boards from different manufacturers to the same bus. This enabled designers to assemble systems with the optimum number and types of boards for the application, and it reduced the software development time required, lowering costs. These are the reasons that board-level products are the basic building blocks used by designers of every type of equipment that uses microcomputers.

Most of these systems conform to bus standards, such as STD BUS, VME BUS, and MULTIBUS. The standards specify the board size, the connector, and the electrical and software requirements to communicate with the other devices attached to the bus. A comparison of bus standards is made in Table 3.1.

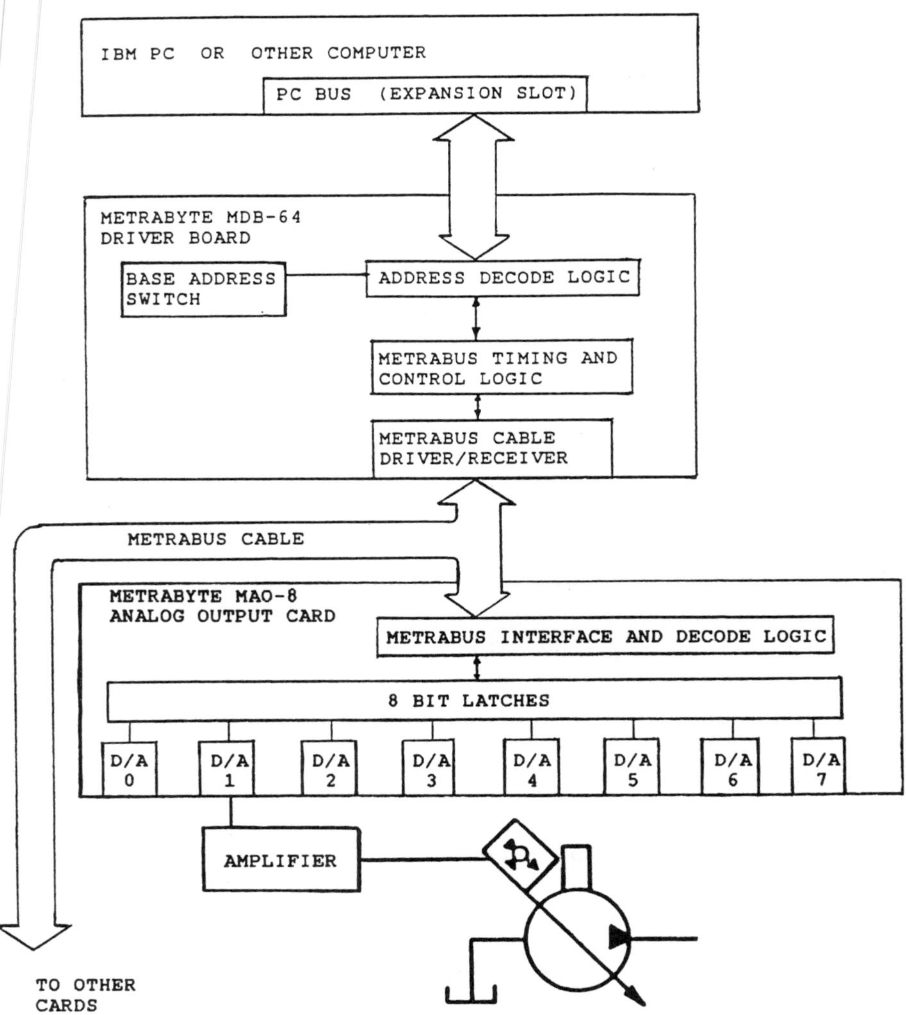

Figure 3.8 Block diagram of Metrabus analog control of an electrohydraulic pump.

The Driver board and MAO-8 board are assigned addresses in statements 10 and 20

10 BASEADR = 768	Driver board is set to an address unused by the PC
20 BRDADR = 16	Sets the address of the MAO-8

Statements 30–80 output a 20 mA and then a 12 mA signal

30 OUT (BASEDAR + 2), 00	Clears the METRABUS
40 OUT (BASEDAR + 1), BRDADR	Sets bus to 16, selecting MAO-8 output channel 0
50 OUT (BASEADR + 1), 255	Output channel 0 set to 20 mA
60 FOR L = 1 to 5000	Remain ON for 5
70 NEXT L	
80 OUT (BASEADR + 1), 128	Output channel 0 set to 12 mA

Figure 3.9 Use of an analog output module to control an electrohydraulic pump.

The STD bus architecture is one of the oldest and is very popular in industrial applications. It is an 8 bit system based on the Zilog Z80 microprocessor. STD bus components are smaller and less costly than components of the other bus standards. Because of this it is better suited to industrial use than the other bus architectures.

The VME bus and MULTIBUS systems are very popular systems aimed at high-performance systems, such as workstations, imaging, robotics, and automation. These systems can be either 16 or 32 bit. They offer greater capability than is needed to control a hydraulic system and are usually too expensive to be used in hydraulic applications.

Generally, board-level microcomputers of the bus standards are not applied in hydraulic systems. The most important reason is the extensive knowledge of software and hardware required to

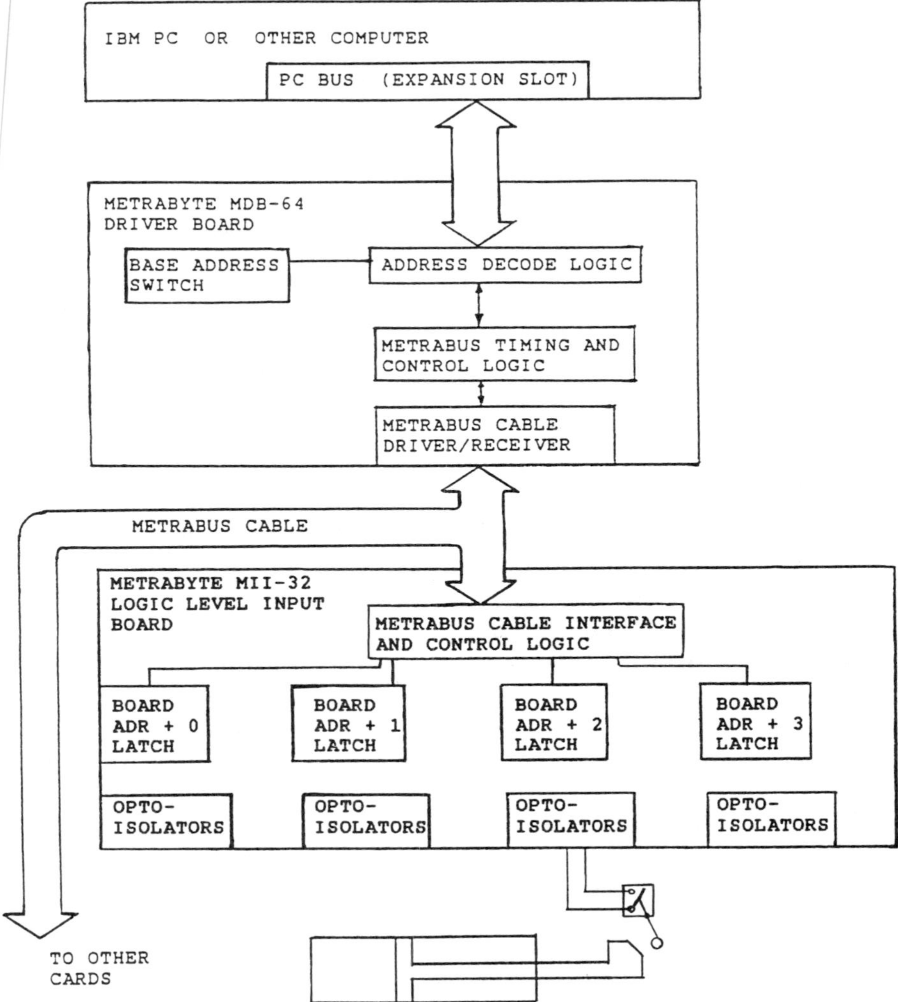

Figure 3.10 Block diagram of Metrabus input module in a control system.

20 BRDADR = 4	Sets the address of the MII-32
30 OUT (BASEADR + 2), 00	Clears the METRABUS
40 OUT (BASEADR + 1), BRDADR	Sets bus to 4, selecting MII-32 address 0
50 DD = INP (BASEADR)	Reads bit 0
60 TEST = DD + 1	Use and strip off all bits except bit 0
70 IF TEST = 0 THEN GO TO 50	If bit is 0 (limit switch contacts are not closed), repeat; (poll address 0 again)
100 REM	

Figure 3.11 Discrete input with a Metrabyte system.

apply these computers. The software for communication between the boards is especially difficult.

Other board-level systems on the market are smaller and more modular and do not conform to bus standards. Some of these have user-friendly software that allows a less experienced person to assemble a system. A manufacturer that markets this type of system is the OPTO 22 company.

This system consists of four types of components (Figure 3.12). The LC series of local controllers are board-level computers capable of controlling a system independently or acting as a slave for a

Table 3.1 Comparison of Three Popular Bus Standards

	STD bus	MULTIBUS	VME bus
Card size (inches)	4.5 × 6.5	6.75 × 12	3.9 × 6.3
Connector type	Card edge	Card edge	Pin in socket
Connector(s)	56 pin	86 and 60 pin	96 and 96 pin
Data bits	8	16	16 or 32
Address bits	16	24	23 or 31
IEEE standard	P961	796	1014

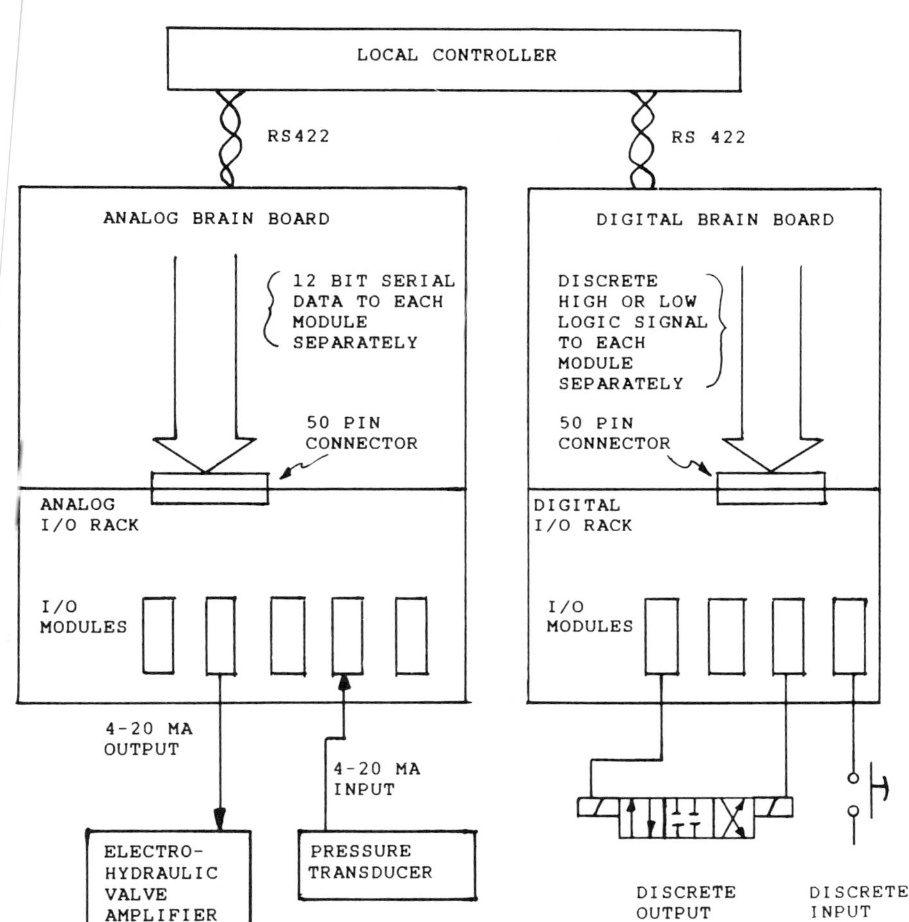

Figure 3.12 OPTO 22 control system with local controller and serial bus.

host computer. The local controllers hold communication software in ROM and a BASIC control program in nonvolatile RAM. On the board is a power supply and battery backup for memory retention. The boards communicate with the I/O boards with a serial RS 422 signal, which passes through a two-wire twisted pair cable. The

BASIC control program can be written by a programmer on an IBM PC and downloaded to the controller through a RS 232 port.

The B series "brain boards" contain a microprocessor and perform serial-to-discrete and discrete-to-serial conversion. The boards receive a RS 422 serial signal from LC controllers or a host computer. The serial signal is decoded, and the specific output corresponding to the received address is activated. This logic-level signal leaves the brain board through a 50 pin connector connected to an I/O rack.

The brain boards can also receive input from the I/O rack. This is accomplished by the brain board reading the logic level (high or low) at one of the pins of the connector.

The B series brain boards are intelligent devices that can perform more functions than serial-to-discrete conversions. Latching, counting, pulse measurement, time delays, and pulsed outputs are possible.

The output modules receive logic voltage signals from the B series brain boards. The logic signals activate solid-state relays in the output modules, capable of switching 115 VAC to power external loads. Isolation is also provided for the logic signal.

The output modules are mounted in racks. Each module is replaceable. A single rack may have as many as 16 discrete output (or input) modules. These racks contain screw terminals for connections to external loads and a 50 pin connector that distributes logic signals to the modules.

Analog I/O is also offered by Opto 22. This is performed by A series brain boards. A RS 422 signal is received from a LC series controller. The brain board decodes the address and data information and opens a communication channel with the addressed output module in the I/O rack. The brain board then sends 12 bit serial data to the analog output module equal to the received data from the controller. The D/A conversion is performed at the output module.

The Opto 22 analog brain boards have another feature. Functions can be performed on these cards rather than at the controller, reducing the software required at the controller. These functions are gain and offset calculations, peak and valley recording, output waveform generation, high and low limit monitoring, and input averaging.

Input can also be performed. An analog input module is required, and it performs the A/D conversion. Serial data are sent to the brain board.

Opto 22 also offers a system that operates with a parallel communication bus, as well as more modules and components than can be discussed in this text. For further information, a company address is provided at the end of the chapter.

3.5 PROGRAMMABLE LOGIC CONTROLLERS

To be useful in an industrial environment, controlling machinery, a microcomputer

1. Must be able to withstand the harsh physical and electrical environment
2. Must have output circuits with sufficient power to drive external electromechanical devices, such as relays and solenoids
3. Must use software that industrial personnel find easy to program
4. Must be serviceable by industrial personnel

Manufacturers of programmable controllers have produced microcomputers that meet these specifications well. Countless variations are available, from many suppliers, to supply a controller that is cost effective for nearly any control task. Let us begin with the early programmable controllers; the relay replacers.

As recently as the 1960s, automated machine control was carried out by electromechanical relays. These racks of dozens of relays contained both the control logic for the system and the power-handling capability to directly drive the solenoids of the directional valves.

This method performed its function but required continual troubleshooting and maintenance as the relays wore and failed. Modifying the duty cycle of a machine or changing the control logic was a time-consuming task because of the rewiring necessary.

Like many obsolete methods, the electromechanical relay was acceptable until an alternative came along: the microprocessor! In the late 1960s General Motors commissioned the Gould Corporation to develop a microprocessor-based controller that would transfer the machine control logic from the relays to software within a microcomputer. This was the birth of the programmable controller.

With the logic in software, reliability was improved because many relays could be eliminated. Changing the logic of the machine control program was made easier because rewiring was no longer necessary. Only the software needed to be changed.

3.6 CONSTRUCTION OF A PROGRAMMABLE LOGIC CONTROLLER

Referring to Figure 3.13, on the top of the PLC is the CPU. It is a microcomputer packaged for an industrial environment. The case is usually impact-resistant plastic sealed against moisture and foreign matter. Electronic suppression and filtration are added to the power supply to prevent voltage surges from entering the digital circuitry. Isolation is used on the inputs and outputs to prevent electrical surges and noise from entering the computer through the I/O devices. The entire system is modular. Each input module and output module and the CPU receive power individually. The only connection between the CPU and the I/O modules is a digital communication bus. With this arrangement, a CPU or a single input or output module can be replaced without disturbing the wiring to the rest of the controller.

Programs are conveniently loaded by a cartridge containing an EPROM (electrically programmable read-only memory). Program development can be done on small, hand-held "terminals," CRT terminals marketed by the controller manufacturer, or personal computers.

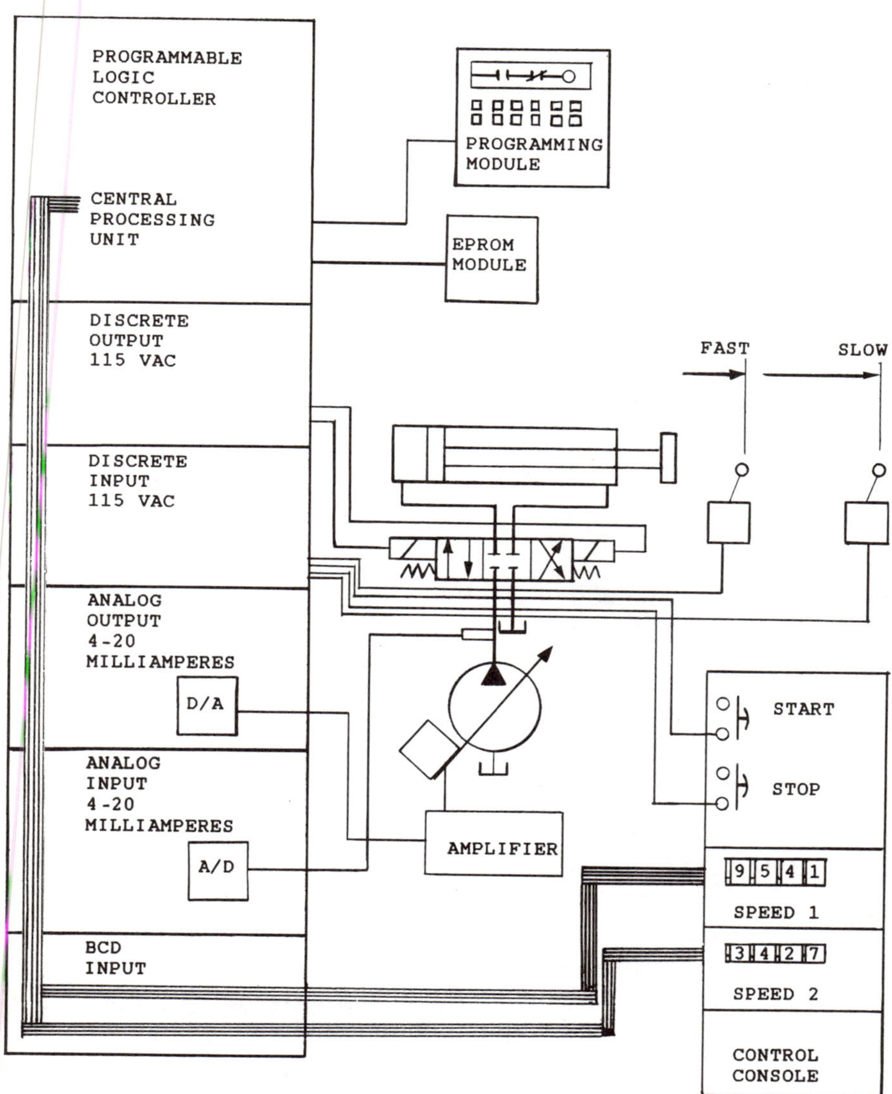

Figure 3.13 A hydraulic system controlled by a programmable logic controller.

Programmable controllers differ from other microcomputers in memory retention. A programmable controller typically has a nonvolatile memory. It retains its program after power to the computer has been turned off. This is important in an industrial environment. Without it, each time a machine was shut off, it would have to be reprogrammed upon restarting. The memory retention is accomplished by nonvolatile memory types or by battery backup.

Relay ladder logic is the most popular language used in programmable controllers owing to the familiarity of maintenance electricians with these diagrams. However, software is continually written that is easier to use and troubleshoot. A discussion of languages is presented in the next chapter.

3.7 PLC-CONTROLLED HYDRAULIC SYSTEM

Figure 3.13 contains a small hydraulic system controlled by a programmable logic controller. The system controls a cylinder that approaches a workpiece in a two-speed sequence. A pump with an electrohydraulic displacement control regulates the speed of the cylinder.

The cycle begins by pressing the start button on the control console. At this point the program was "waiting" for an input to the discrete input module.

As the input was received, the relay ladder logic program began its execution of the duty cycle by activating the discrete output that controls the 115 VAC power to the extend solenoid of the directional valve. At the same time the "speed 1" analog output is activated, which sends a 19.27 mAmp signal to the pump's amplifier card.

$$19.27 = [.9541 \times (20-4)] + 4$$

The cylinder rod moves at high speed (95% pump output flow) until it contacts the first limit switch. The limit switch sends a 115 VAC signal to the discrete input module. The program receives a signal, deactivates the speed 1 analog output and activates the speed two-analog output. The controller now sends a 9.48 mA signal to the amplifier card, resulting in 34% flow from the pump.

The cylinder rod moves at a reduced speed until it contacts the second limit switch. The switch sends a 115 VAC signal to the discrete input module. The program then receives a signal that causes it to deactivate the analog output, which sends a 4 mAmp signal to the amplifier card, resulting in 0 flow. The discrete output to the extend solenoid of the valve is also deactivated.

If the shifting of the valve is delayed until the pump has reached 0 flow, the stopping of the cylinder is very smooth, without shock to the machine. The program then activates the discrete output to the retract solenoid of the valve. At the same time, the analog out speed 1 is activated to retract the cylinder rod at high speed.

If the operator desired to change either speed setting, he or she could do so by changing the numbers on the BCD thumbwheel input switches.

The program can be written to sense the activation of the first limit switch on the retract stroke to slow the retract as it approaches the end of the cylinder stroke. Another limit switch can be added at the end of the stroke, or a timer can be programmed operating from the actuation of the limit switches to signal the program that the sequence is finished.

There is no selector switch on the control panel for repeating the cycle. Therefore the sequence must be programmed to cycle indefinitely, or cycle until a counter in the program accumulates a specific number of cycles, or cycle once and stop.

3.8 CLASSES OF PLCs

PLCs can be classified into three groups based upon the number of I/O and cost:

1. Large (200 2000 I/O), over $5000
2. Medium (50 200 I/O), $2000–$5000
3. Small (less than 50 I/O), $200–$1500

3.8.1 Large PLCs

These are used for the control of very large systems or in automation applications in which they send commands to local controllers. The features of large PLCs are as follows:

2000 discrete and analog input and output points
Documentation (report generation)
Higher level languages (other than relay ladder logic)
More communication capability than smaller PLCs
Greater memory size
BCD input
PID loop control
Motion control
Arithmetic capability

3.8.2 Medium PLCs

The medium class of PLC was used in the hydraulic system example. These controllers have many of the features of the larger controllers but are smaller in size, cost, and memory. These controllers also vary widely in capability and can be configured into a cost-effective solution for a wide range of applications. The capabilities of medium-sized PLCs differ significantly between makes and models, and care should be taken in evaluating controllers in this group. Particular care should be taken if calculation capability is desired. Calculations as well as other microcomputer functions can be difficult or impossible in some models of medium-sized PLCs.

3.8.3 Small PLCs

These units are designed for low cost. The only I/O available is discrete. These controllers are designed to replace electromechanical relays and are cost effective if the number of relays exceeds five. These units provide sequence control equal in reliability to the larger controllers.

VENDOR SOURCE LIST

Board level computers and structured I/O, Opto-22, 15461 Springdale Street, Huntington Beach, CA 92649.

Control interfaces for the IBM PC, Metrabyte Corporation, 440 Myles Standish Boulevard, Taunton, MA 02780.

PLCs, Gould Modicon, G.E., Allen Bradley, Texas Instruments, Reliance Electric; Square D, Omron, and many others.

BIBLIOGRAPHY

Control Engineering Magazine, 875 Third Avenue, New York, NY 10022.

Programmable Controls, 67 Alexander Drive, P. O. Box 12277, Research Triangle Park, NC 27709.

4

Microcomputer Programming Languages

4.1 INTRODUCTION

Microprocessor-controlled hydraulic systems require software, like any systems controlled in this manner. The sequence of events and decision-making logic must be entered into the computer. Nearly any computer language can be adapted to machine control, but a few have risen to prominence for a number of reasons.

In this chapter, three "levels" of software are discussed: dedicated software, user languages, and programmer languages. Most of the discussion centers on the user language of relay ladder logic (RLL). The reason for this is the popularity of RLL and the practical nature of this language for machine control. The language is easy to learn, and the hardware to run it is available and economical. In addition, many of the principles discussed in the programming of RLL are typical of machine control tasks that must be carried out in any computer-controlled machine cycle.

The term "dedicated software" refers to software resident in a machine, which the user does not change. The user typically replies to prompts, or questions from the computer, and the machine makes adjustments to the programmed duty cycle. The program itself is not altered. With a system such as this, programming knowledge is not required.

Programmer languages are the languages used to write dedicated software and user languages. These are mentioned only briefly in this book. Instruction in the application of assembly code or high-level languages, such as C, goes beyond the scope of this book. These languages are mentioned to provide an overview of the software tools available.

4.2 RELAY LADDER LOGIC

Before microprocessors were invented, automated machinery, especially in the automobile industry, was controlled by electromechanical logic. The machine sequence and limited decision making were performed by complex networks of electromechanical relays. These performed the intended function but required maintenance as the relay contacts or solenoid coils wore. Maintenance often required lengthy and expensive troubleshooting in complex systems. Changes to the logic could also be time consuming and expensive.

Changes to the logic were so difficult for the automobile industry, during new model changeovers, that racks of relays were often discarded entirely rather than attempt reprogramming. It was because of this that General Motors, in the late 1960s, commissioned the Gould Corporation to develop a microprocessor-based controller that contained the control logic in software rather than hardware. The result was the programmable controller.

Relay ladder logic is a carryover from the electromechanical era. This language is popular now because of the many maintenance electricians who are comfortable with it. Enhancements have been added to this language that allow the user to go beyond discrete (on or off) tasks to simplistic computer functions, such as timing, counting, and arithmetic.

Because of the popularity of this language, a short course in relay ladder logic is presented here. The principles discussed here are valid for most makes of programmable controllers, but there are differences in each manufacturer's versions of RLL. Programming manuals from each manufacturer must be consulted.

4.2.1 Programming in Relay Ladder Logic

RLL is a symbolic language that represents an electrical circuit. The "electric current" of the logic diagram is thought to flow from left to right (Figure 4.1).

The circuit or RLL diagram of Figure 4.2 contains two rungs, a toggle switch, two control relays, and a set of contacts. Before throwing the switch, no "current" is flowing through either rung. The control relays CR 1 and CR 2 are not activated because of the open circuit of the open switch in the top rung, and the open contacts (CR 1) in the second rung.

As the switch is closed, current flows to the CR l control relay. This activates the contacts CR 1, allowing current to flow through the bottom rung and activating the control relay CR 2.

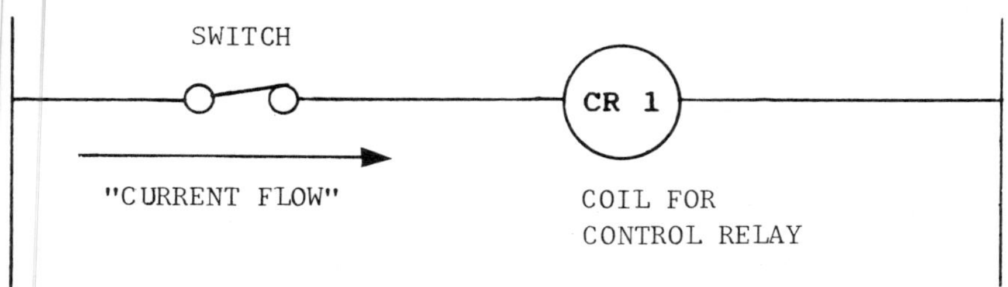

Figure 4.1 A program line in relay ladder logic.

In Figure 4.3, the control relay CR 2 sends a signal out of the programmable controller to the outside world. In this case a lamp is lit.

An important construction in RLL is that of a latching relay (Figure 4.4). The purpose of the latching relay is to begin a machine cycle from a momentary switch closure and continue the cycle until stopped by another device. Without a latching device, the operator would have to hold the switch closed throughout the cycle.

As the pushbutton is pressed, current flows to CR 1. This activates the contacts CR 1. The contacts then hold the top run on.

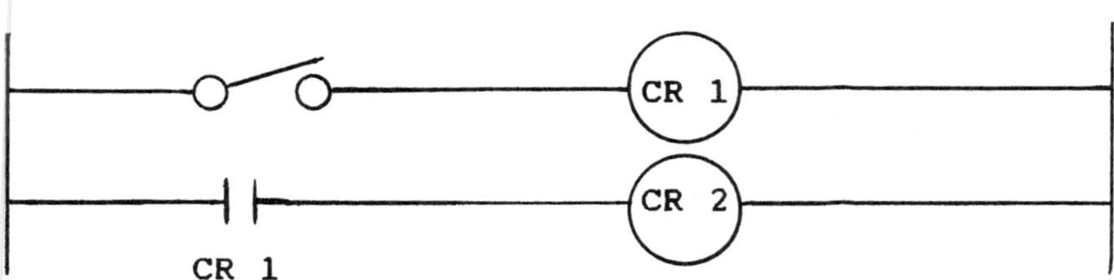

Figure 4.2 A program in relay ladder logic.

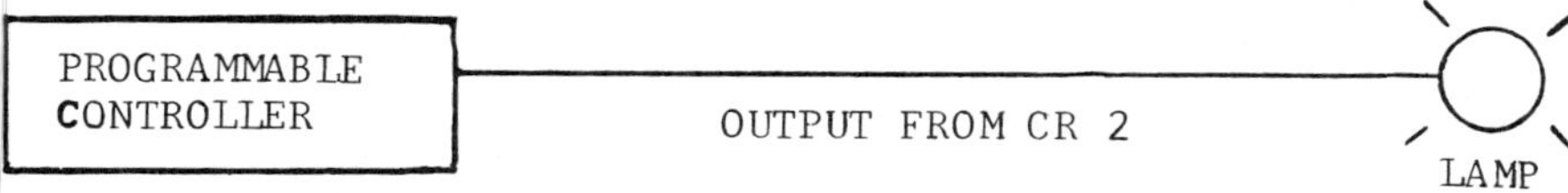

Figure 4.3 The hardware between the program and the machine element to be controlled.

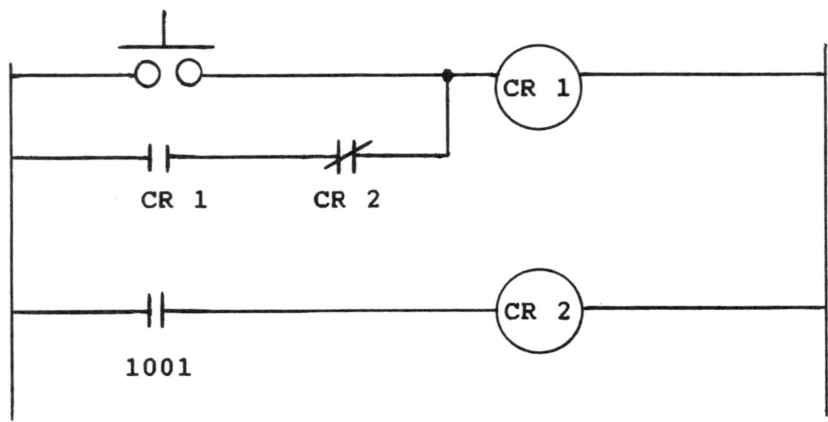

Figure 4.4 Use of a latching relay.

The rung would remain on indefinitely if it were not for the bottom rung and the limit switch LS 1 in Figure 4.5. As the cam on the machine pushes LS 1, a signal is received by the input module of the programmable controller. The RLL program contains the LS l limit switch in the form of an input register 1001. This means that the program continually scans the register 1001. As the limit switch signal is sensed by the program, the LS 1 symbol of the second rung is activated. Current then flows to CR 2, activating it. CR 2 then activates the contacts CR 2, which open the circuit or rung 1. This deactivates CR 1, stopping the machine cycle and opening the contacts CR 1.

The important concepts to remember are the *latching relay*, the *input register 1001,* and the use of the *normally closed contacts* of CR 2.

4.2.2 TIMERS

Timers are function blocks added to RLL that emulate the function of mechanical or electrical timers. Shown in Figure 4.6 is a rung with a timing block.

PROGRAMMABLE
CONTROLLER
INPUT 1001
LS 1
MACHINE

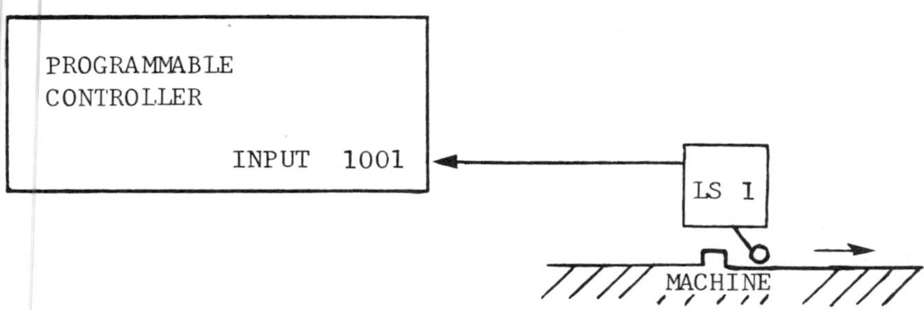

Figure 4.5 Limit switch.

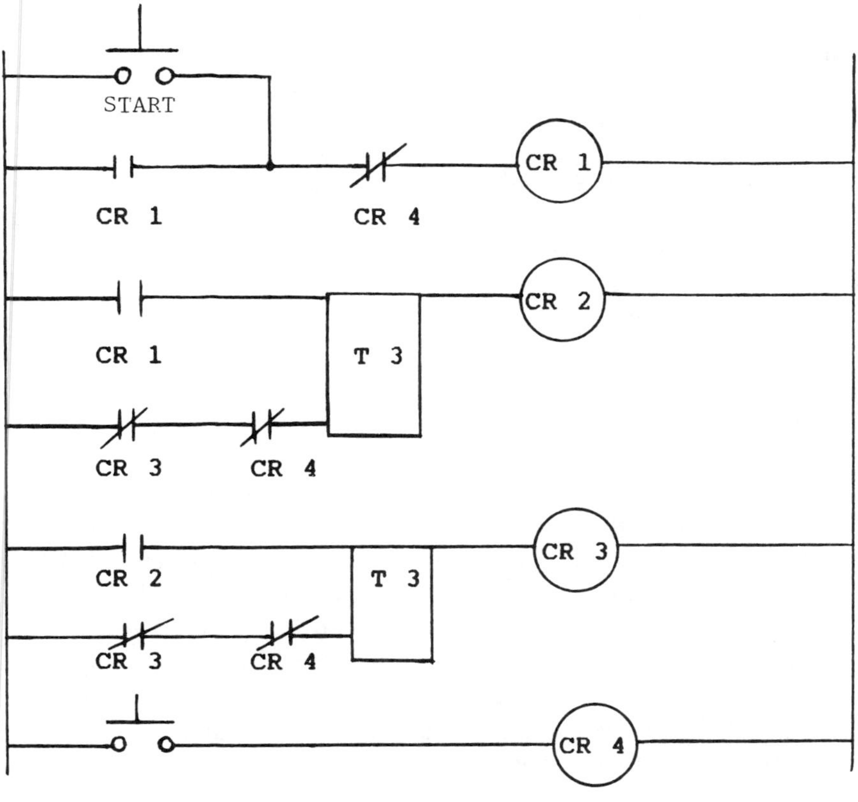

START
CR 1
CR 4
CR 1
CR 1
CR 2
T 3
CR 3
CR 4
CR 2
T 3
CR 3
CR 3
CR 4
CR 4

Figure 4.6 RLL program with timing blocks.

As the switch is activated, the timer begins its cycle, which in this example takes 3 s. After 3 s the timer activates its output, which activates CR 2. CR 2 now remains on indefinitely, until the signal to the lower part of the block is interrupted, by activating either CR 3 or CR 4. Activating CR 3 or CR 4 opens the normally closed contacts, opening the circuit, which resets the timer.

Machine duty cycles typically contain many sequential events. Many valves and other components perform a given function after one action occurs and before another. This can be seen in Figure 4.7, a cycle diagram for a typical machine.

Timers are useful in programming sequences and in ensuring that actuators operate in sequence, not at the same time. Figure 4.6 contains the RLL diagram that produces the sequence in Figure 4.7.

Another useful construction is a *repeating circuit.* In the RLL diagram of Figure 4.6, the process begins with a momentary push-button contact, activates CR 2 after 3 s, holds CR 2 on for 3 s and off for 3 s and then on, repeating continuously until deactivated by the switch. Note that the stop switch activating CR resets the timers and unlatches the top rung containing CR 1.

4.2.3 Counters

Counters are another function block added to RLL that emulate electromechanical counters. The counter counts the number of times it receives a signal to its input and outputs a signal when the programmed count is reached. In Figure 4.8, a counter can be added to the diagram of Figure 4.6 to stop a process after three cycles. At a count of 3 a signal is sent to CR 5, which stops the process and resets the timers. The counter can also be connected to an external sensor, such as a photoelectric switch, to count the passing of objects on a conveyor before stopping the process.

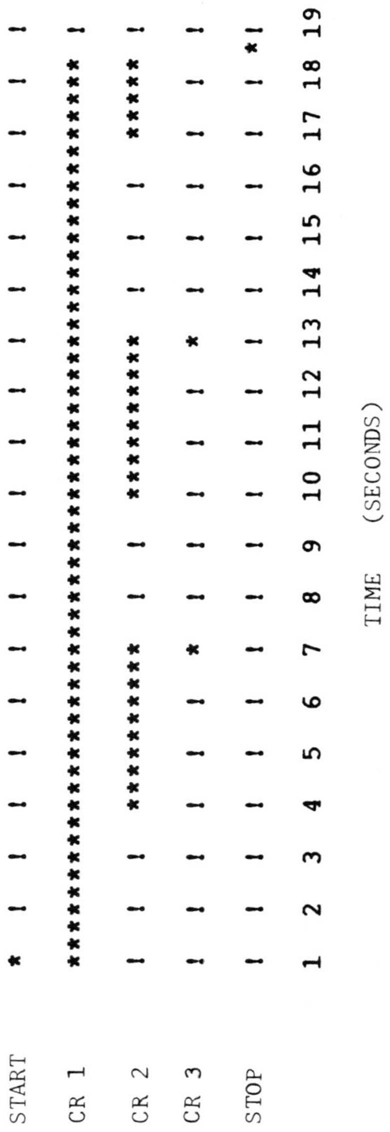

```
START    *  !  !  !  !  !  !  !  !  !  !  !  !  !  !  !  !  !  !
CR 1     *****************************************************  !
CR 2     !  !  !  **********  !  !  **********  !  !  !  *****  !
CR 3     !  !  !  !  !  !  *  !  !  !  !  !  *  !  !  !  !  !  !
STOP     !  !  !  !  !  !  !  !  !  !  !  !  !  !  !  !  !  ! *!
         1  2  3  4  5  6  7  8  9 10 11 12 13 14 15 16 17 18 19

                        TIME   (SECONDS)
```

Figure 4.7 Cycle diagram for a machine.

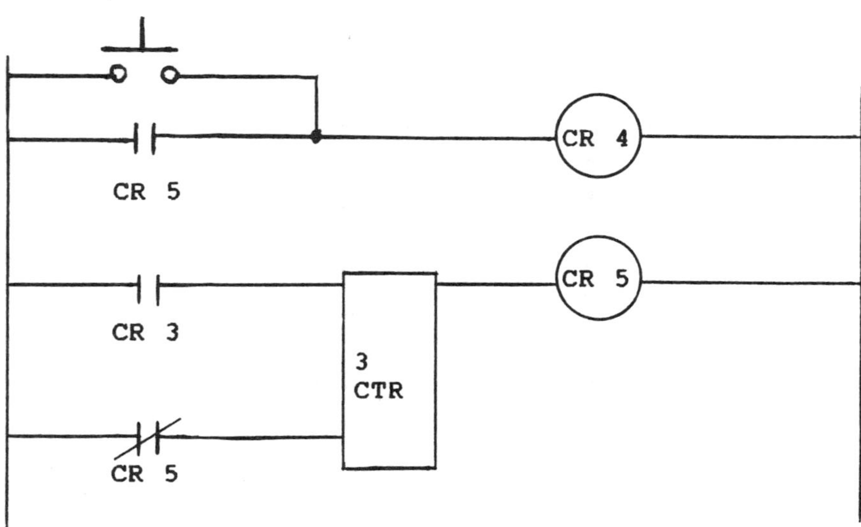

Figure 4.8 RLL program with a counter block.

4.2.4 Analog Output Function Blocks

Analog signals, typically 0–5 V, can be sent out of a programmable controller by activating the function block. Figure 4.9 contains a function block with the register 3001. Upon closing the switch, the contents of the register are sent to the external device, in this example a proportional valve amplifier card.

The register contents, the number 500, are obviously not in the voltage range of 0–5 V. Typically, in programmable controllers the numbers in registers are between 0 and 1000. The number 500 would result in an output voltage proportional to the register contents divided by 1000. Therefore the output voltage sent from the PC would be $(500/1000) \times 5 = 2.5$ V. The register contents 500 could be programmed or controlled by external devices. It could also be the result of arithmetic calculations, of arithmetic function blocks.

The important point is that the function block, with the analog output module, can be programmed to output an analog (or proportional, in hydraulic terms) signal to a hydraulic pump or valve.

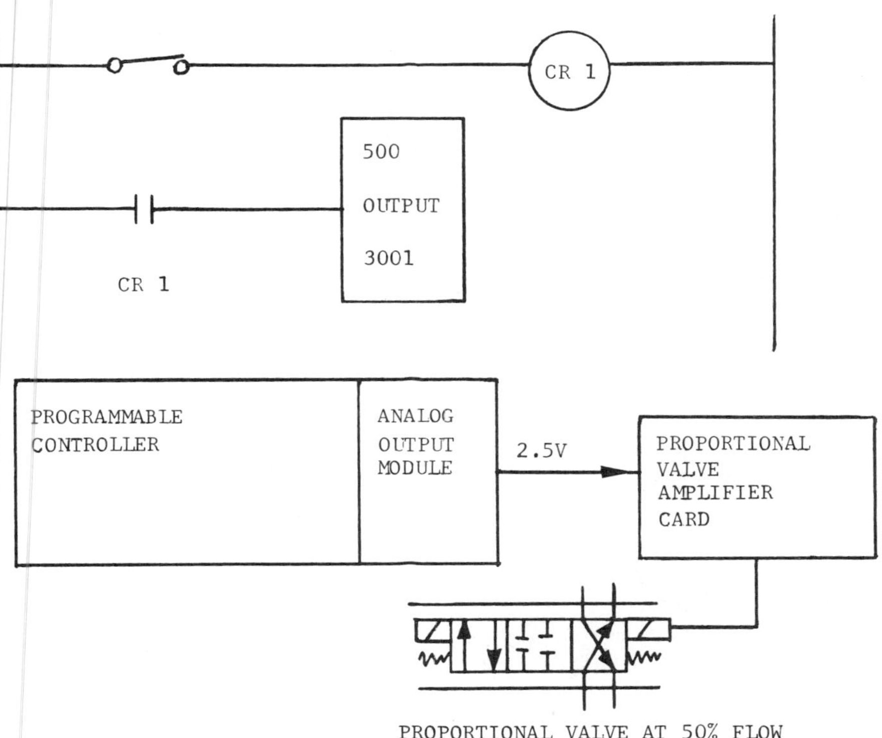

Figure 4.9 Analog output function block controlling a proportional valve.

4.2.5 Analog Input Function Blocks

Analog information can also be received via a function block of the RLL program. The analog information can be from a pressure transducer, as shown in Figure 4.10.

The pressure transducer contains circuitry that scales a pressure from 0 to 2000 psi to a voltage from 0 to 5 V. In the example, the PC is programmed to execute a function controlled by coil 7 when the pressure reaches 1000 psi. As 1000 psi is attained, the pressure transducer sends a 2.5 V signal to the analog input module. The 2.5 V signal is then computed to a fraction of 1000 and

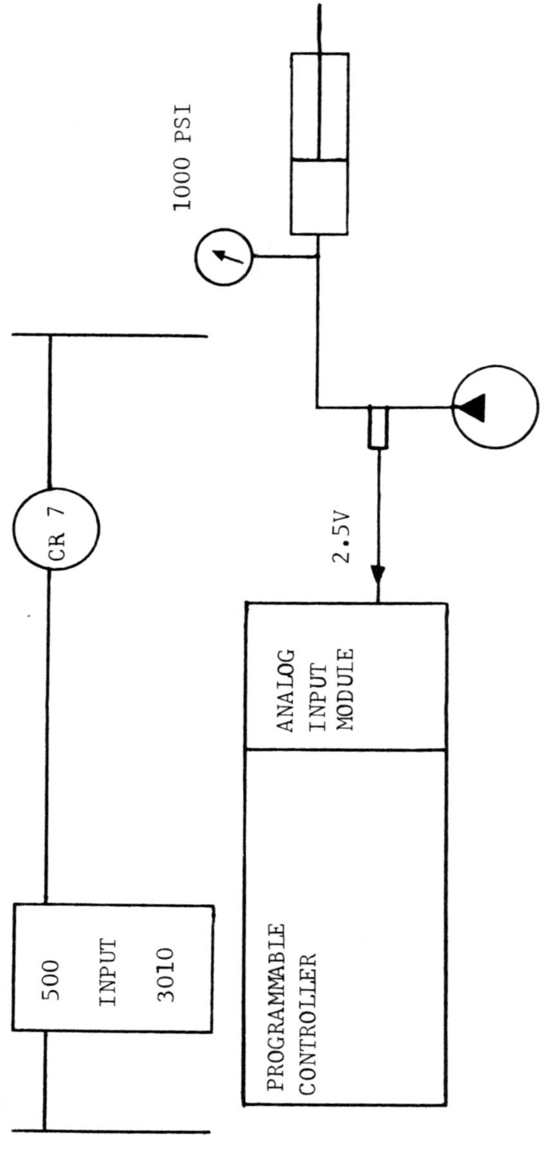

Figure 4.10 Input of sensor valves in RLL.

compared to the register contents. The calculation (2.5/5) × 1000 is 500. Because this number is equal or greater to the programmed register contents, the input function block activates the coil 7.

Therefore, an analog input function block receives an analog voltage, compares it to the contents of a register, and activates a relay depending upon the comparison.

4.3 MANUFACTURER-SUPPLIED PROGRAMMING LANGUAGES

Some manufacturers of programmable controllers have written their own programming languages. These languages are commonly based on English language statements and are quite easy to understand and program with. An example of one of these languages is MCL from Cincinnati-Milacron. Following is a listing of a MCL program to sequence traffic lights. Note the ease of understanding what is occurring in the program.

```
REPEAT
  SET    (MAIN-GREEN, SIDE-RED) ON
  DELAY 6 SEC
  SET    (MAIN-GREEN, MAIN-YELLOW) INV
  DELAY 4 SEC
  SET    (MAIN-YELLOW, SIDE-RED) OFF
  SET    (MAIN-RED, SIDE-GREEN) ON
  DELAY 6 SEC
  SET    (SIDE-YELLOW, SIDE-GREEN) INV
  DELAY 4 SEC
  SET    (MAIN-RED, SIDE-YELLOW) OFF
LOOP
END CYCLE
```

4.4 DEDICATED SOFTWARE

Some controller manufacturers have written software packages that enable the user to program the machine sequence without using a programming language. The controller presents the user with a series of menus and prompts (questions). The user simply types in answers to the questions, and the controller does the programming. This is the easiest to use of the programming methods presented, since no programming knowledge is required. The only drawback is availability. Dedicated software is costly to develop and therefore is available only in machines with significant quantities in operation.

Programmable motion controllers often use this method. The user is prompted for locations, directions, and speeds. The controller then uses this information to control step motors or other devices to position the head of a machine tool.

4.5 BASIC

Developed in 1962, its name is an abbreviation for beginner's all-purpose symbolic instructional code. This is the most popular language used in personal computers, but there are few applications of this language in machine control. A brief discussion is presented here for comparison with other languages since many people are familiar with it. Although not the ideal machine control language, it is a convenient language to use in experimentation with personal computers and interfacing.

In BASIC, interfacing to the outside world is accomplished by the two statements PEEK and POKE. The PEEK statement can be used to examine the contents of a memory location of the computer. The BASIC program can then use the number from this location in a calculation and alter the results of the program. If this memory location is connected to the outside world, such as a parallel port, an external electronic circuit can be used to communicate a number from the real world to a running computer program.

The POKE statement is an output statement. By *poking* a number to the memory location of a parallel port, a voltage is induced at specific pins of the output port. The pins receive a voltage depending upon the binary (base 2) equivalent of the number the program sends to the port.

A sample program to turn on a row of light-emitting diodes, is presented in Figure 4.11. The external circuit is also shown. The computer used is a Commodore 64. The program counts from 0 to 255, showing the numbers in binary, with the LEDs. The reader should be aware that PCs differ in memory location numbers, BASIC statements, and the methods of interfacing to external circuits.

BASIC is an easy-to-use language with the capability to do almost anything. It is seldom used in machine control applications for the following reasons.

Lack of execution speed is a characteristic of interpreter BASIC. With the interpreter, each line of the program must be decoded and executed one by one.

A compiler can be used to speed up the execution of a BASIC program. The compiler translates the entire BASIC program into machine code and stores this machine language program. The interpreter is not used. The drawback of this technique is the difficulty of editing and listing the program. Once compiled, a program cannot be "uncompiled" from machine code back into BASIC. Therefore the convenient editing and troubleshooting of the interpreter BASIC is lost. The original BASIC program must be rewritten and compiled again.

Getting back to interpreter BASIC, which is far more popular than the compiled version, another drawback of BASIC in machine control is understanding or reading the code. BASIC is not a graphic or symbolic language like relay ladder logic. BASIC does not have the symbolic "rungs of a ladder" that perform like electrical circuits. However, this is probably not the most serious drawback. Some popular machine control languages, such as C and Assembler, are also less than graphic.

A reason that BASIC is not used may be that it is an "in-between" language in terms of capability and user convenience. It

```
Counting Program

10   POKE 56579, 255
20   FOR X = 1 TO 255
30   H = 255 -X
40   POKE 56577,H
45   FOR V = 1 TO 100        (delay loop)
46   NEXT V
50   NEXT X
90   POKE 56579, 0           (turns off port)
```

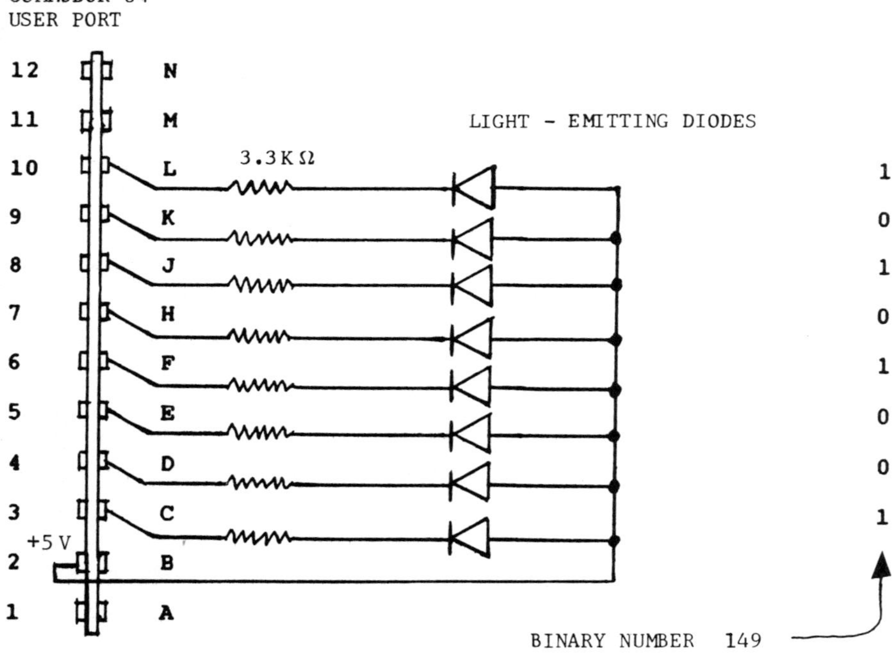

Figure 4.11 Control using BASIC in a personal computer.

does not have the speed of such languages as C or Assembler. It also lacks the readability and editing convenience of relay ladder logic or manufacturers' languages used in a machine control situation.

4.6 C

C is a structured programming language used mainly by programmers writing dedicated software packages. It has the speed of execution approaching assembly code, with powerful statements that result in an economical code. The drawbacks for its use in hydraulic system control are its difficulty of use and the few people who can program using it. A machine sequence would be time consuming to write and difficult to change or modify in the field. C is mentioned here to acquaint the reader with a high-level language used to write dedicated software for some of the most sophisticated control applications.

4.7 ASSEMBLY CODE

Assembly language is the lowest level language one can program with. It is the language closest to machine code. It offers the programmer the most direct and therefore fastest execution time of any language. However, drawbacks to the use of assembly language prevent its use in most hydraulic and other applications:

It is extremely time consuming and difficult to program with because of the number of statements required. It may take 50 or 100 lines of assembly code to accomplish what 2 lines of BASIC can do. As with all computer languages, the time and effort required in programming increases with the number of statements in the program.

Few people know assembly code. This is an important consideration if the user of the machine would ever desire to modify the program.

In general, assembly code is used by experienced programmers who must have the greatest speed of execution possible. For hydraulic applications, this level of speed (less than .010 response time) is seldom needed. The systems that require such speed require programming knowledge and microprocessor expertise beyond the scope of this text.

4.8 SPEED

Speed of execution is dependent more upon software than hardware. For most hydraulic control applications, the speed of response with any of the languages presented here is adequate. Speed can become a problem in closed-loop systems in which the computer is closing the loop. The rule of thumb is that assembly code is required to close the loop if the required response time is .010 s. Programmable controllers vary in response time from .01 to .06 s.

An important point is that if a system requires a response time of .060 s or less, from the time a signal enters the controller to the time a control signal exits the controller, the designer should closely examine the specifications of the controllers under consideration. A controller with a scan time of .040 s cannot respond in less than .040 s *unless* the events are programmed in an efficient manner (adjacent rungs of a ladder for input and output, for example).

Speed of execution can be influenced by the organization of the program. In a relay ladder logic program, the speed of program execution may be dependent upon the scan time of the controller. If the timing of a certain function is critical, the input and output rungs should be short and close together in the program.

4.9 PROGRAMMING TIPS

These are a number of programming steps that one can follow in programming a sequence for a hydraulic system or other machine.

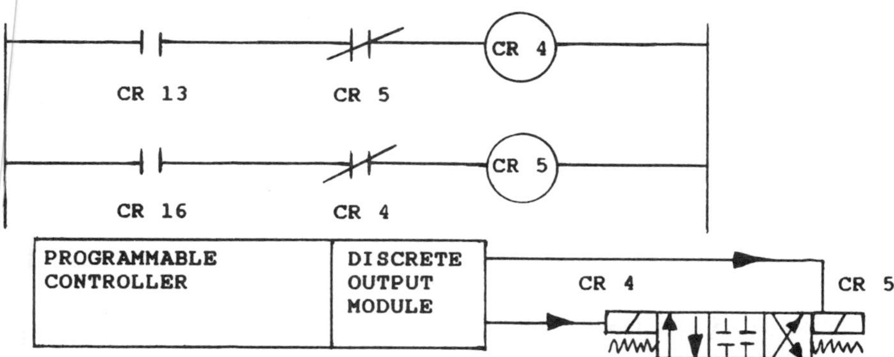

Figure 4.12 Example of good programming practice in RLL.

1. Make a timing diagram of the cycle.
2. Note which events occur at the same time. Is it desirable or necessary that these events occur at the same time? If it is, then tie them to the same output of the controller. If it is not necessary, connect the devices to different outputs.
3. Test the program logic as thoroughly as possible before operating the machine. This saves time in determining if a problem is caused by software or hardware.
4. Identify events that are time dependent versus sequence dependent. If an event is sequential (it must follow another event), it should be programmed sequentially. Programming sequential events based upon time can result in the disruption of the sequence, in the case of a delay. It is also easier to change the cycle without logic errors if the program is sequential instead of time dependent.
5. Identify events or outputs that must be prevented from ever being energized at the same time. A good idea is to add a program statement that prevents this from occurring, as shown in Figure 4.12.

Step 5 is redundant, but it can save a costly failure in the event of an errant software modification or a logic error in the software that results from an unexpected combination of inputs. A practical example is the two-solenoid hydraulic directional valve in Figure 4.12. If both solenoids were energized at the same time, neither would move. The current through the solenoids would remain at inrush current levels. This current is great enough to exceed the ratings of the outputs of some programmable controllers.

4.10 PROGRAMMING STYLE

Modularity is important. Groups of instructions, or rungs of a ladder, that control a specific portion of the sequence should be grouped closely together. Jumping from place to place in a program makes it more difficult to troubleshoot and modify.

4.11 SUMMARY

In selecting a microprocessor-based controller to control a hydraulic system, the first item to consider is the software that will be used to program it.

Software considerations should include the expertise of the programmers who do the initial programming and the expertise of the people who troubleshoot or modify the software after it is in operation.

If programming is avoided, the purchase of a machine with dedicated software (menu driven) should be considered.

If a machine with dedicated software is not available, controllers with relay ladder logic or manufacturers' languages (MCL) may be needed. Relay ladder logic or MCL is easy to learn and use. Consultants with expertise in RLL are numerous.

5

Output from the Microcomputer

5.1 INTRODUCTION

In Chapter 2 the output from a microprocessor was discussed, which took us to the output port of a microcomputer. In this chapter, the use of the data from the output port is discussed.

Information travels out of a microcomputer through output ports of two types, serial and parallel (Figure 5.1). Discussed in this chapter is the hardware required to transform these signals into control signals for hydraulic components.

Another method of interfacing to a microcomputer was discussed in Chapter 3—the addition of an interface card to the bus (via an expansion slot) of a microcomputer. The output port tasks are then handled by the interface card. This method is not repeated in this chapter, to give more attention to microcomputer output ports.

5.2 SERIAL OUTPUT

Serial data output is a single stream of data bits, sent one at a time, over a two-wire cable (Figures 5.1, 5.2, and 5.3). A high bit, or logic one is sent as a voltage pulse of -5 to -15 V and a low bit, or logic zero is sent as a pulse between $+5$ and $+15$ V (according to the popular RS 232 standard). The speed of data transmission is 300–19,200 bits/s (RS 232).

Other standards for serial data transmission have different voltages for the logic levels and transmission speeds (see Table 5.1). The voltage levels for digital logic circuitry also differ. Discussions of logic levels appear later in this chapter.

Serial data transmission is not as fast as parallel data transmission because it is one bit at a time instead of one data word. Serial has the important advantage of requiring only two wires to transmit data. Parallel data transmission requires one for each bit, in addition to control lines.

The advantage of the two wires in serial transmission is a small concern in lengths of less than 100 feet but represents a great savings in wiring costs in lengths of hundreds or thousands of feet.

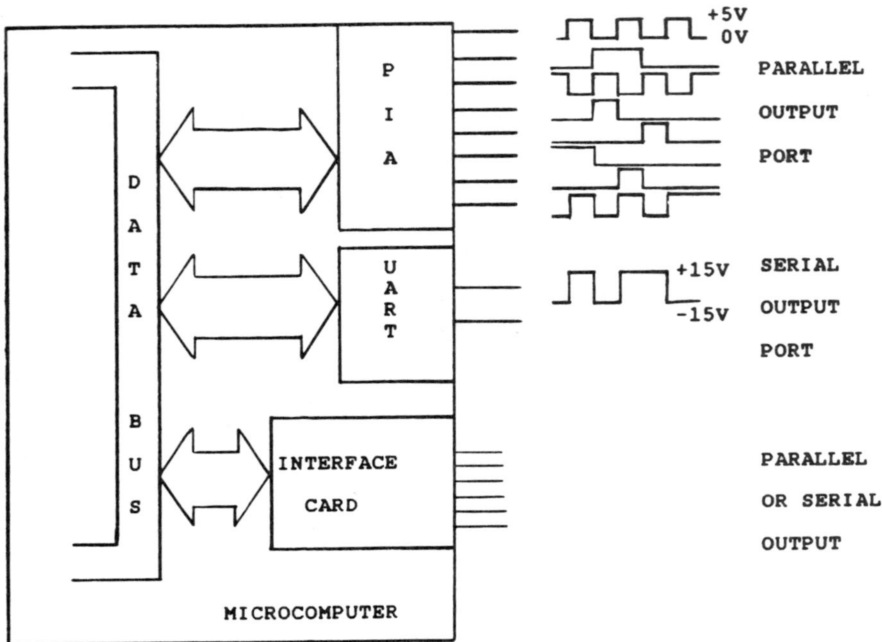

Figure 5.1 Output from a microcomputer.

Serial transmission also makes it practical to send data over conventional telephone lines. It is not practical to send data over eight separate phone lines. This is required for 8 bit parallel transmission.

5.3 SERIAL DATA TRANSMISSION STANDARDS

For reliable serial transmission, the timing and voltage level of the pulses are important. Standards have been written that specify these parameters.

RS 232 is the most widely used data communication standard for short distances and moderate data rates. The standard specifies both the voltage levels for the logic 1 and 0 signals and the function of each of the pins at the interface.

Table 5.1 Comparison of Serial Data Standards

	RS 232 C	RS 422	RS 485
Voltage levels	Single-ended	Differential	Differential
Signal voltage			
Maximum	±15	±5	± 5
Minimum	± 5	±2	±1.5
Maximum cable length (feet)	50	4000	4000
Maximum data rate (bits/sec)	20,000	10 million	10 million
Number of drivers on one line	1	1	32
Number of receivers on one line	1	10	32

Figure 5.2 shows the voltage ranges for the transmission and reception of data according to the RS 232 standard. From −5 to −3 V and from +5 to +3 V is a margin that allows a voltage drop of 2 V between the transmitter and the receiver. Therefore a logic 0 signal transmitted at a minimum level (+5 V) arriving at the receiver at 3 V is accepted as a logic 0. The region between −3 and +3 V is an invalid region.

RS 232 is a single-ended voltage standard. A logic 1 is dependent upon its voltage potential with respect to ground. Logic 0 is also dependent upon its potential with respect to ground. If both the 1 and 0 signals were to receive interference of +15 V, it is obvious that many of the logic 1s would be interpreted as logic 0s.

Other transmission standards are highly tolerant of interference because the logic levels are dependent upon the difference between the signals, not on an absolute point in the circuit. RS 422 is one of these differential standards. It is highly immune to electrical interference because of this.

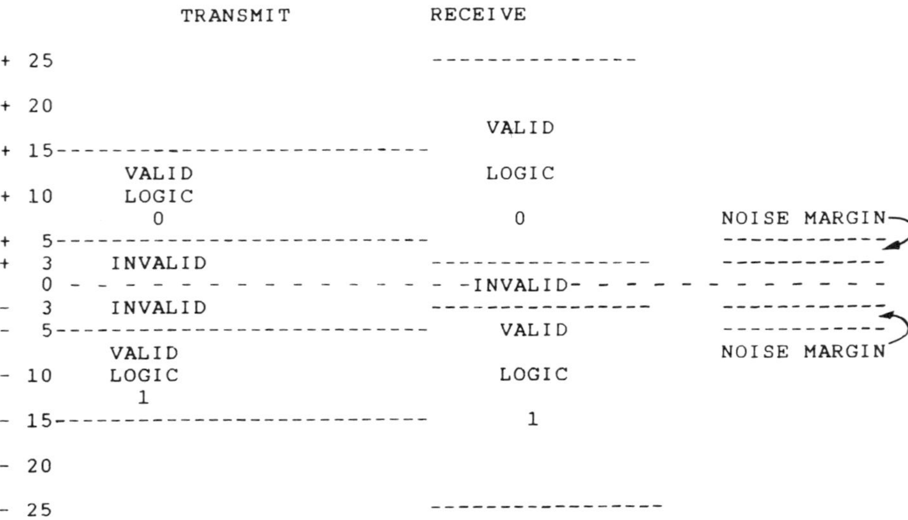

Figure 5.2 RS 232 voltage standards.

Figure 5.3 shows a data transmission circuit using the RS 422 standard. Any voltage on wire A that is greater than the voltage on wire B is interpreted as a logic 1. The driver could be an AMD (Advanced Micro Devices) 26LS31 and the receiver a 26LS322.

RS 422 is replacing RS 232 in industrial applications because of the longer cable lengths possible and the greater electrical noise immunity. Conversion components are available for converting the signal of one standard to that of another.

The RS 232 standard is ambiguous in some of the pin assignments. Connection of two noncompatible RS 232 devices does not

Figure 5.3 RS 422 serial data transmission.

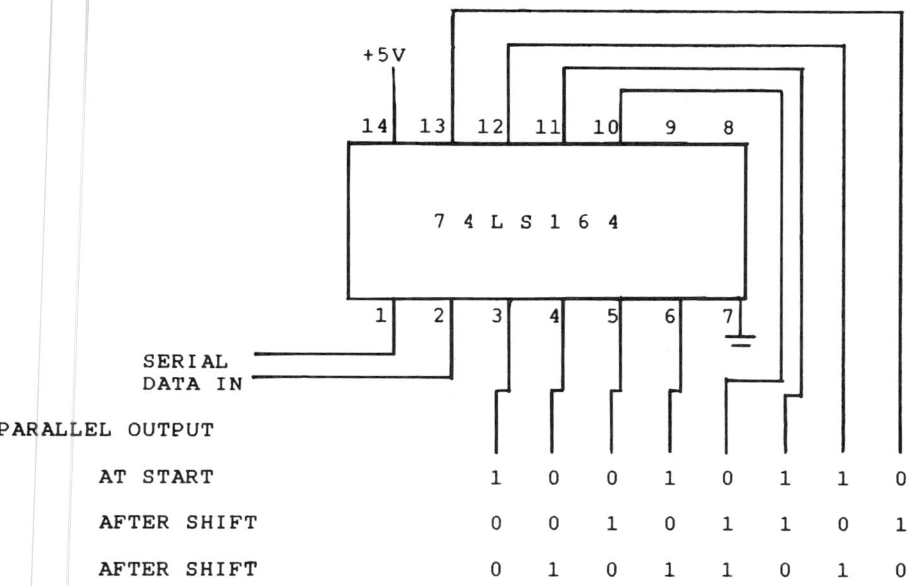

Figure 5.4 Function of a shift register.

function and may damage the equipment. Most of the RS 232 noncompatibility problems can be solved through cable modifications. A useful tool for this is a gender changer, which converts a female RS 232 to a male end. “Smart” cables are also available to make the conversions. Many of us have already experienced these compatibility problems when attempting to connect a printer to a personal computer.

5.4 SERIAL-TO-PARALLEL CONVERSION

The data on the microprocessor's data bus are in parallel form and must be converted into serial form for serial transmission. This is done with shift registers and latches (as explained in Chapter 2). After transmission, the serial data must be converted back into parallel data for an external device to make use of the data. This is also done by shift registers and latches (Figure 5.4). *Note*: Driver

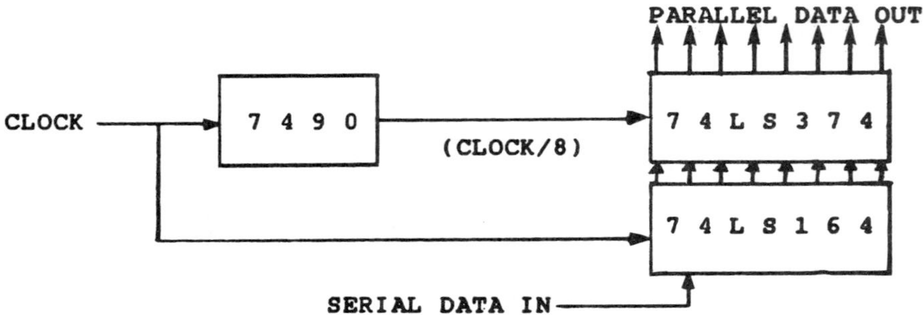

Figure 5.5 Serial-to-parallel conversion.

ICs may also be required to convert voltage levels from one standard to another.

The 74LS164 is a shift register that can be used to convert serial to parallel data. It operates by receiving serial data, 1 bit at a time, storing the bit, and advancing the bits through eight memory locations. For example, if the binary number

10010110

were in the memory of the 74LS164 and another bit was received, a high bit, the parallel output pins would be changed to the number

00101101

by shifting the bits over one location. If another bit were received, a low bit, the number would become

01011010

The serial-to-parallel conversion is not accomplished by the shift register alone. Another chip, a latch (Figure 5.5), must receive the parallel output from the shift register and store it. Furthermore, it must receive it at the correct time, the time at which the shift register contains one byte of data, with each bit in the correct location. The least significant bit (LSB) must be in the LSB location, and the most significant bit in its location at the other end of the data word.

At this instant in time, the 74LS374 latch receives the parallel byte of data from the shift register and holds it. The timing of the data transfer is done by the 7490 chip. It divides the clock pulse by 8 and sends this signal to the latch. Upon receiving this signal, the latch opens its input to the shift register to receive the byte of data.

The byte of data can be held indefinitely or until the next byte is ready in the shift register. The 74LS374 is used as a parallel output port to hold and send a parallel data byte to external devices, such as digital-to-analog (D/A) converters or displays.

5.5 PARALLEL OUTPUT PORT

A parallel output port has a single pin for each data bit. It can output an entire data word (a binary number) at once. The port (PIA, parallel interface adaptor) can hold a binary number while waiting for an external device to use the data for another operation (D/A conversion, for example). A popular 8 bit parallel standard is the IEEE 488.

5.5.1 Interfacing to a Parallel Port

Here we discuss interfacing external devices to a parallel port, but this also applies to a system in which the serial output from a microcomputer has been converted from serial to parallel by an external component. There are two methods to extract meaningful data from a parallel output port: the use of the binary number at the output port and the logic state (high or low) at one or more of the output pins.

5.5.2 Use of the Pin States for Discrete Output

A simple method of obtaining a number of discrete outputs equal to the number of data bits of the parallel port is shown in Figure 5.6. Each data output pin drives a transistor that controls the current flow through an external load. In this circuit a logic low at the PIA allows a current of 2 mA to flow into the PIA. The PIA is

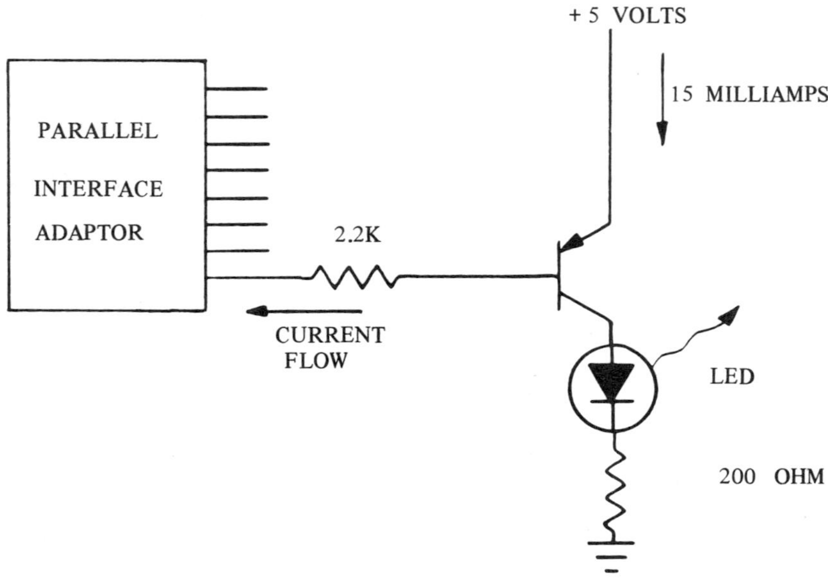

Figure 5.6 Switching on an LED with a single pin.

"sinking" the current. A logic high would switch off the LED (light-emitting diode).

A circuit can also be constructed that switches current on through a load with a logic high output from the PIA. This is shown in Figure 5.7. This circuit also shows another concept: driving a larger load with a logic output. The two transistors are arranged in a "Darlington" configuration. The same NMOS PIA from Figure 5.6 could drive a load of 2 A (with a gain of 1000 Darlington transistor) or a load of 20 A (with a gain of 10,000 Darlington).

In order to drive high current loads, several techniques are available in addition to a Darlington. Relays can be used, solid-state or electromechanical. Solid-state relays have the advantage of speed and no moving parts to wear out. Electromechanical relays have the advantages of zero leakage current and greater current capacity. Electromechanical relays can handle current flow in both directions, and some solid-state devices can handle

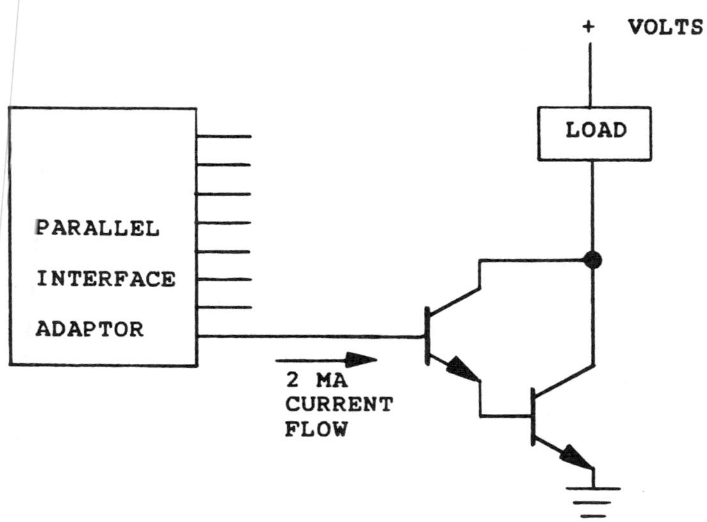

Figure 5.7 A single pin driving a Darlington transistor.

current in only one direction. Power transistors, SCRs (silicon-controlled rectifiers), and triacs are examples of solid-state relays.

A technique used to switch higher voltages with digital logic is to use an open collector device. In Figure 5.8 a circuit is shown in

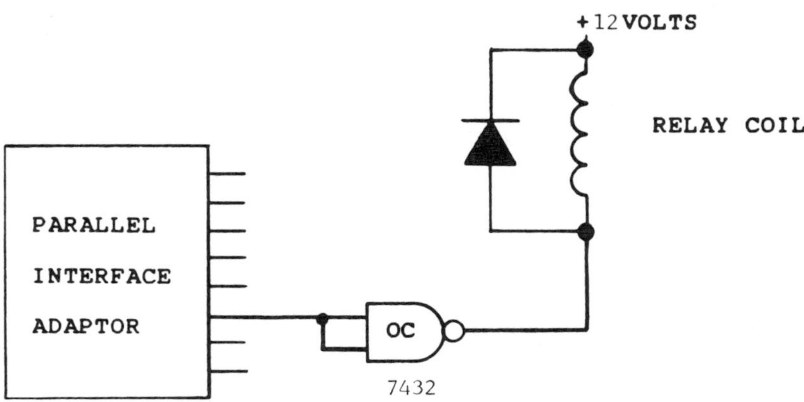

Figure 5.8 A single pin driving an electromechanical relay.

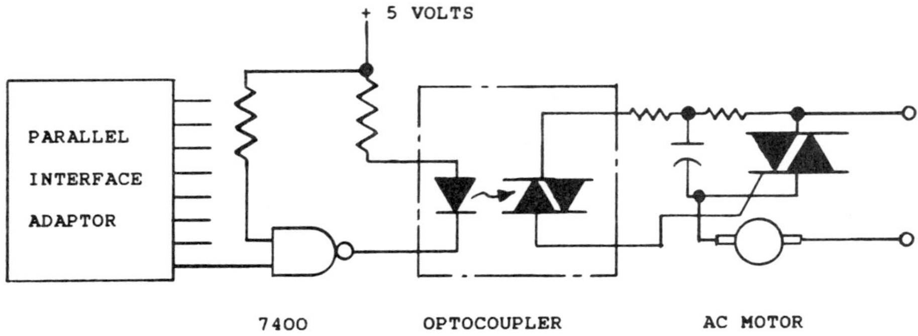

Figure 5.9 Switching AC loads with an optocoupler and triac.

which a PIA and a 7432 driver (with an open collector) is used to switch on a 12 V electromechanical relay. A diode is used to bypass the inductive current generated by the relay coil. Open collector devices are available with voltage ratings as high as 80 V (DS3611–3614).

Shown in Figure 5.9 is a circuit used for switching a 115 VAC load with a PIA. A 7400 buffer powers an optocoupler, which switches on a triac.

The buffer is a 7400 NAND gate. A buffer is often used between a PIA and a load to provide more current capacity for driving external loads. A buffer also provides some electrical isolation between the load and the digital circuitry.

The optocoupler provides electrical isolation between the external load and the digital circuitry. Isolation between external circuitry and the microcomputer's logic circuitry is always a good idea. Voltage transients or static electricity can cause errors in the logic signals of the microcomputer or damage to the IC chips. The most common form of isolation is by optical means, also referred to as an optoisolator.

An optoisolator or optocoupler is a combination of an LED and a phototransistor. The light from the LED turns on the transistor. Because there is no electrical connection between the digital control circuit and the external load, voltage transients cannot travel from the power circuit back into the control circuit.

A triac is a common solid-state relay made by connecting two SCRs in parallel. Triacs can handle current flow in both directions. They are commonly used to switch ac currents controlled by digital circuitry.

Discrete digital logic chips are often interfaced to PIA output port. These logic components (e.g., AND gates, OR gates, and flip-flops) are constructed using technology different from that of the NMOS PIAs, probably TTL or CMOS. Care must be taken that the voltage threshold levels are compatible between the transmitter and the receiver. The lowest logic 1 voltage from the transmitter must exceed the logic 1 threshold of the receiver.

Care must also be taken to see that the resistance downstream of the output pin limits current from the PIA to its rating. If the current out of the PIA is too great, the voltage may drop at the PIA to below the logic threshold of the receiver.

5.5.3 Conversion Between TTL and CMOS

The logic levels of TTL circuits differ from the levels of CMOS and NMOS circuits. Microcomputer circuits using CMOS logic must often be interfaced to discrete logic components, which may be either CMOS or TTL. Figure 5.10 shows the logic levels of TTL. CMOS logic is dependent upon the power supply used. The logic threshold is at one-half the power supply voltage. Therefore a 2.4 V high signal from a TTL device is not interpreted as a logic high. A pull-up resistor must be added from the CMOS input to the positive supply. This raises the TTL output above the CMOS threshold of 2.5 V. An alternative to the pull-up resistor is the use of 74HCT and 74ACT CMOS devices, which are TTL compatible.

5.5.4 Use of a Binary Number at the Parallel Port for Discrete Output

Using the binary number at the parallel port, digital logic circuits can be used to decode the number and activate an output, as shown in Figure 5.11. The output is activated if the digital number transmitted from the port is 10011000.

```
                TRANSMIT                 RECEIVE

5 VOLTS ---------------------   ------------------------------

4         LOGIC                                   LOGIC
          HIGH                                    HIGH

3

2.4 -------------------------------
                        NOISE MARGIN
2                         -----------------------------------  2

           INVALID                                INVALID
1
                                -----------------------------0.8
           LOGIC LOW                              LOGIC LOW
0 VOLTS - - - - - - - - - - - - -  - -  - - - - - - - - - - - -
```

Figure 5.10 TTL logic levels.

5.6 ANALOG OUTPUT

Many external devices that receive commands from a microcomputer require an analog voltage signal. (Amplifiers for electrohydraulic valves, for example, require an analog command voltage.) This analog voltage can be any voltage between 0 and 5 V, proportional to a binary number in the computer. An example is an analog output of 3.5 V. This is a 70% command signal.

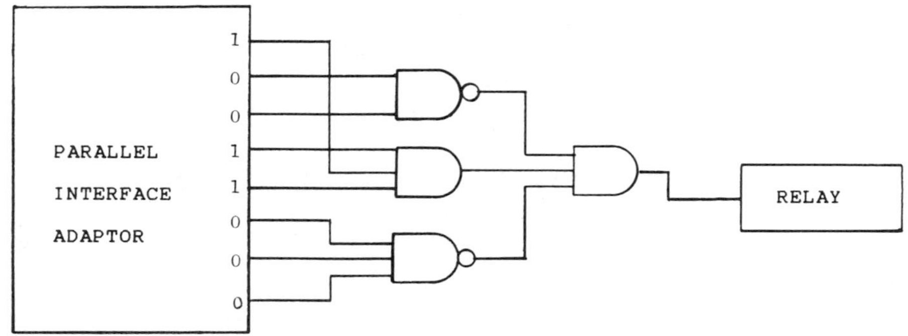

Figure 5.11 Decoding with digital IC chips.

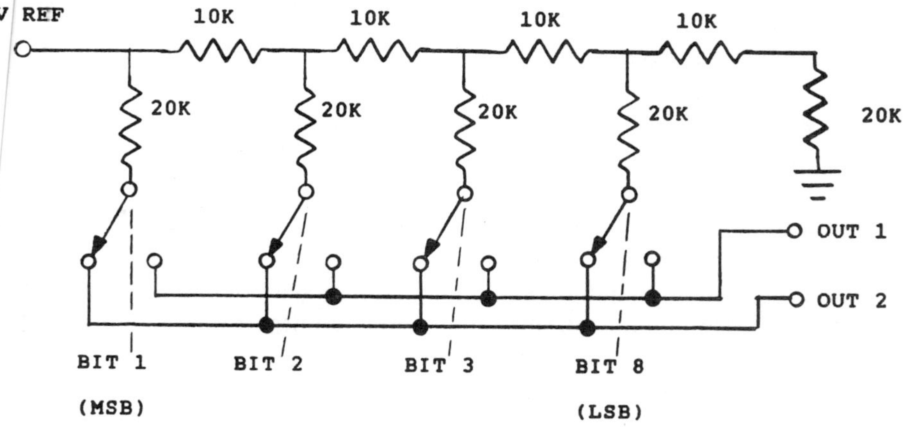

Figure 5.12 Digital-to-analog converter.

Another popular analog signal is a 4–20 mA current loop. A current between 4 and 20 mA is the output, and it is also proportional to a binary number in the computer. Using the same 70% command signal, the analog output current is 15.2 mA. The 4–20 mA standard is common in the process industries and has a high resistance to electrical interference.

Analog output is obtained from a parallel port by a digital-to-analog converter. D/A converters are quite simple in design, consisting of a resistor network. (Figure 5.12). The output is a sum of the currents switched by the bits. By applying different voltages to V^{ref}, the D/A converter becomes a multiplying device.

Analog output can be obtained from serial output by serial-to-parallel conversion followed by D/A conversion. Analog Devices manufactures a chip that performs the entire serial-to-analog conversion, the AD 7543.

5.7 SCALING OF ANALOG OUTPUT

The output from a D/A converter may not match the minimum and maximum values required for an application. We can compensate somewhat through programming, but then resolution of the output signal may be reduced.

Obtaining a reduced voltage range from a D/A converter or any signal source can be done with a voltage divider circuit. Figure 5.13 contains a circuit that converts the 0–5 V range from the converter to a 0–3 V range. This circuit has a potential problem of nonlinearity. Current flow out of the 0–3 V output affects the voltage division between the two resistors, resulting in a nonlinear voltage conversion. This problem can be reduced to an insignificant level if the resistance downstream of the 0–3 V output is 10 or 100 times the resistance of either of the resistors in the voltage divider.

If a voltage divider circuit cannot be used, an operational amplifier circuit can be constructed, as shown in Figure 5.14. This is referred to as an attenuator. It has a gain of 1/10.

Obtaining a larger voltage range than the converter output is typically done with an operational amplifier circuit, as shown in Figure 5.15. The gain of the circuit is determined by the ratio of the resistors:

$$\text{Gain} = 1 + \left(\frac{R_1}{R_2}\right)$$

$$V_{output} = V_{input}\ (1 + R_1\ R_2)$$

Therefore if R_1 = 20 KΩ and R_2 = 50 KΩ, then the gain is 2.5. A 0–5 V range from the converter becomes 0–12.5 V.

In addition to changing the gain of a circuit, it may be desirable to add an offset. This adds a voltage to all the output voltage values. In Figure 5.16 an operational amplifier is added to the circuit of Figure 5.14 to change the voltage range from 0–12.5 to 1–13.5 V. Bias is another term for offset.

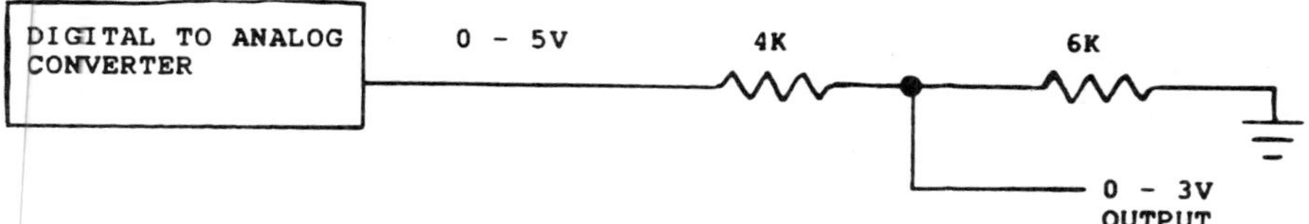

Figure 5.13 Voltage divider circuit.

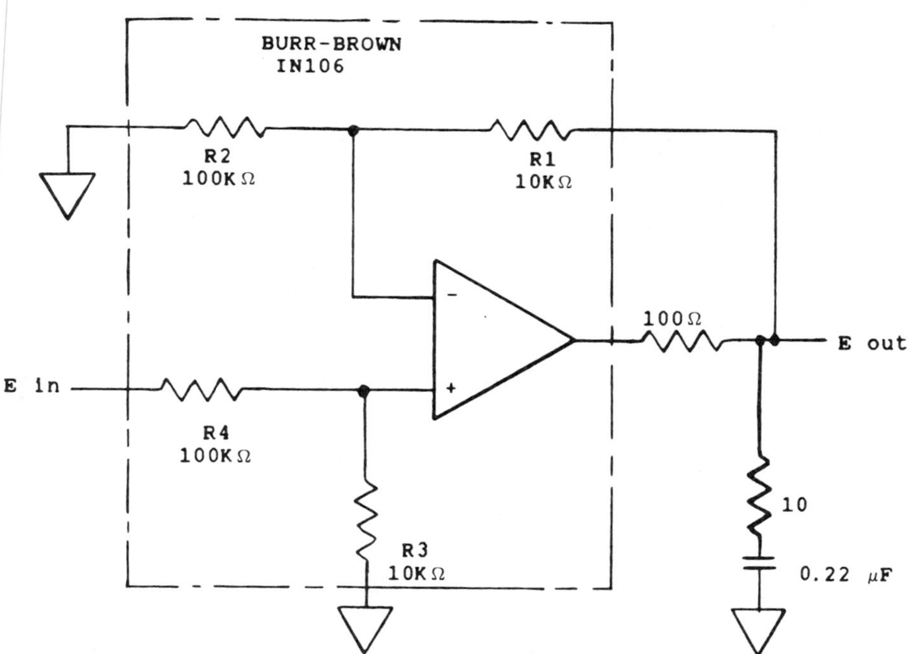

Figure 5.14 Gain of 1/10 attenuator.

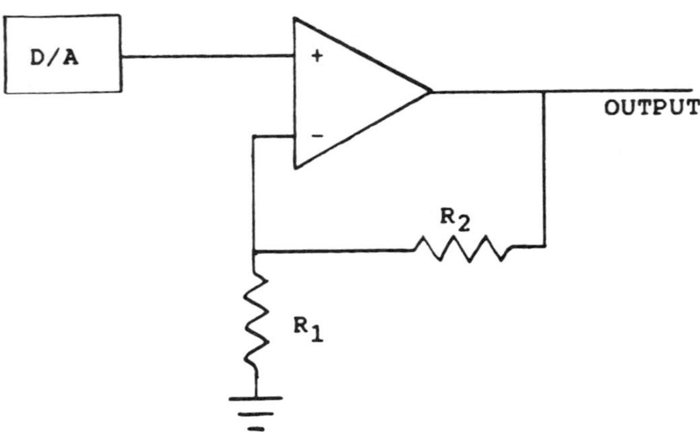

Figure 5.15 Expanding signal range with an operational amplifier.

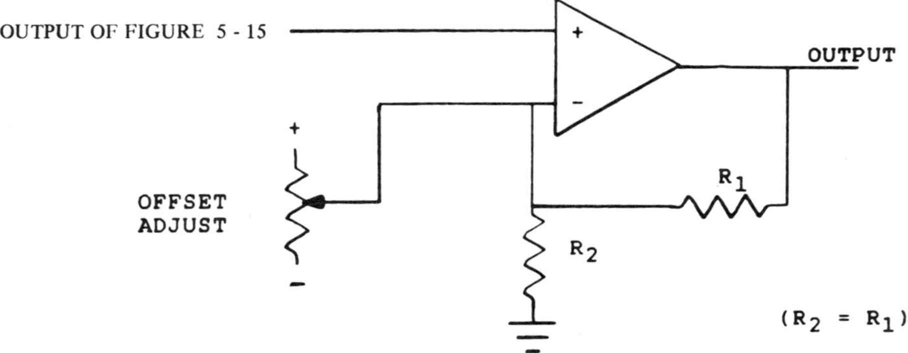

Figure 5.16 Voltage offset with an operational amplifier.

VENDOR SOURCE

Burr-Brown Corporation, International Airport Industrial Park, P. O. Box 11400, Tucson, AZ 85734.

6

Input into the Microcomputer

6.1 INTRODUCTION

A microcomputer that controls a hydraulic system must accept signals from the machine operator. In most systems the hydraulic system itself also sends signals to the microcomputer, which must be accepted. These signals can be of various forms and usually must be converted into parallel data for the input port (PIA) of the microcomputer to accept. In this chapter, the conversion of many types of control and sensor signals into parallel data for the PIA is discussed. Also discussed are control signals that do not pass through the PIA; hardware interrupts.

6.2 DISCRETE INPUT

The simplest input into the microcomputer is that of a switch closure—discrete input. This is an on or off signal from start switches and limit switches. If the system to be controlled has fewer switch inputs than the number of data lines of the parallel port, a simple interface circuit, as shown in Figure 6.1, can be used. The PIA receives a logic high with a switch open. (The PIA contains a pull-up resistor.) A closed switch grounds the input. To receive the logic value, the PIA must be commanded by the central processing unit (CPU) to be in a receive mode.

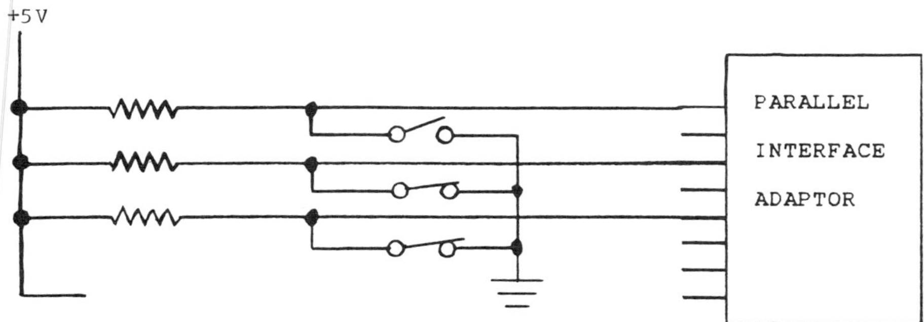

Figure 6.1 Discrete single pin input.

If the number of input switches in a system exceeds the number of input lines of the PIA, digital logic ICs can be used to code the input into a multiple-bit binary number. Figure 6.2 shows a circuit and truth table for inputting six discrete inputs to three lines of a PIA. A disadvantage of this approach is that only one input can be high at a time. If multiple inputs are high the input to the PIA is ambiguous. Different combinations of open switches result in the same input to the PIA.

To ensure that the PIA receives a unique value for each switch opening, a priority must be established. Figure 6.3 shows a truth table for a Fairchild 9318 priority encoder. Notice that the highest number input that is low sets the output of the 9318. Lesser number inputs are ignored.

In many systems, "switch bounce" can be a problem. This occurs as a switch is opened or closed and consists of a number of

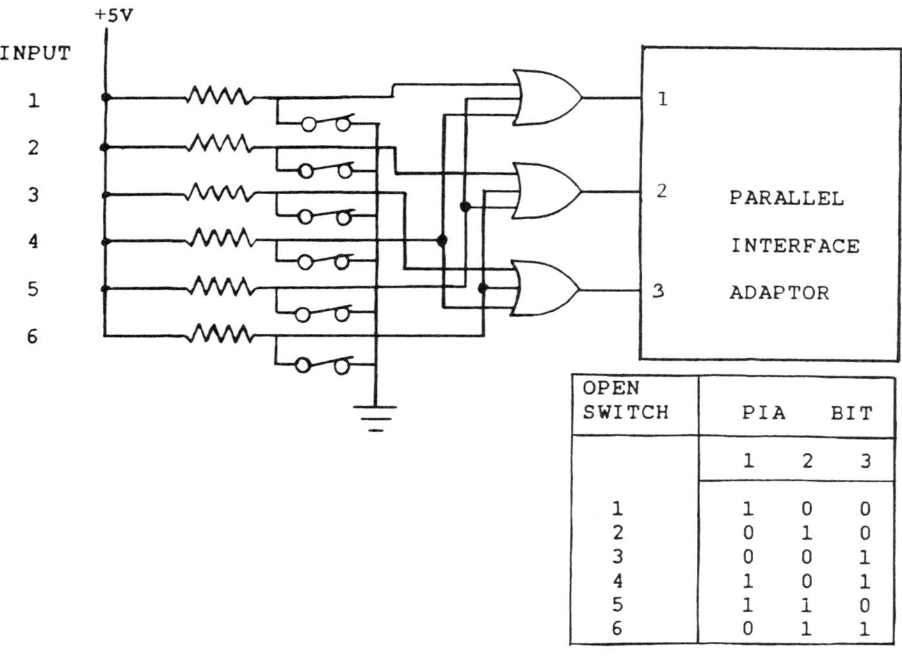

OPEN SWITCH	PIA BIT		
	1	2	3
1	1	0	0
2	0	1	0
3	0	0	1
4	1	0	1
5	1	1	0
6	0	1	1

Figure 6.2 Encoding multiple discrete inputs.

INPUT									OUTPUT				
E	0	1	2	3	4	5	6	7	G_S	A_0	A_1	A_2	E_0
H	X	X	X	X	X	X	X	X	H	H	H	H	H
L	H	H	H	H	H	H	H	H	H	H	H	H	L
L	X	X	X	X	X	X	X	L	L	L	L	L	H
L	X	X	X	X	X	X	L	H	L	H	L	L	H
L	X	X	X	X	X	L	H	H	L	L	H	L	H
L	X	X	X	X	L	H	H	H	L	H	H	L	H
L	X	X	X	L	H	H	H	H	L	L	L	H	H
L	X	X	L	H	H	H	H	H	L	H	L	H	H
L	X	L	H	H	H	H	H	H	L	L	H	H	H
L	L	H	H	H	H	H	H	H	L	H	H	H	H

Figure 6.3 Truth table for 9318 encoder. H = high-voltage level; L = low-voltage level; X = don't care.

high and low signals, which the computer can interpret as multiple switch cycles. One way to avoid this problem is in the software. Be sure that the program responds to the first switch action that it receives, not any following closures in a given time period. If a bounceless switch is required, the circuit shown in Figure 6.4 prevents this problem. Other techniques are also used [1].

In many systems, it may be necessary to interface high-voltage signals to a microcomputer. An example is a 115 VAC signal from a pressure switch. The 115 VAC must be isolated from the PIA to prevent damage to the computer. This is accomplished by the circuit shown in Figure 6.5. An optically coupled solid-state relay, a Teledyne C67AI-1, provides the isolation. This relay is packaged in a 8 pin DIP configuration.

6.3 POLLING

Getting the logic input to the pin of the PIA is part of the task of input of a switch closure into a microcomputer. The PIA must hold

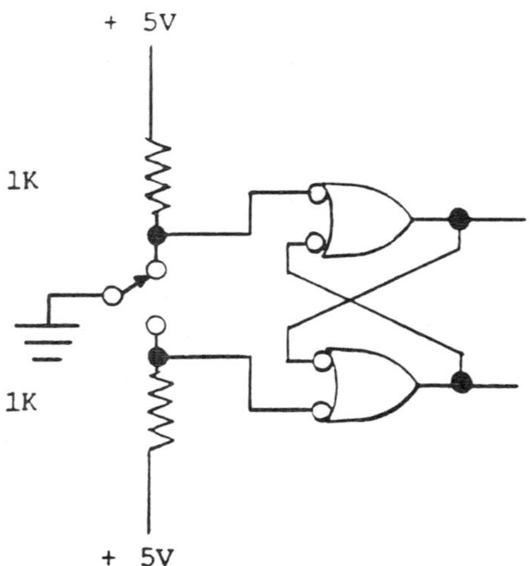

Figure 6.4 Switch debouncing circuit.

the data (a binary number) for a period of time, and the microprocessor must then ask the PIA for this number. After the microprocessor receives this number and notices that a switch closure (or opening) has taken place, it either begins to execute its machine control program or does a branching operation and executes a different section of the computer program.

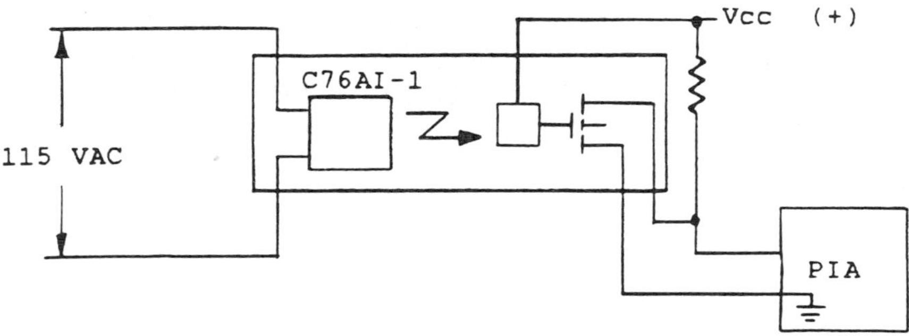

Figure 6.5 Converting high voltage to a logic input.

How do the CPU and the PIA know in advance that a switch closure will occur? Advance knowledge is required for the PIA to input the data and hold it at the proper time for the CPU to call for it. This is done by polling. The CPU, at regular intervals (50 times per second, for example), signals the PIA to input data from the parallel port and hold it, with a signal on the control bus (Figure 6.6). A short time later (nanoseconds or microseconds) the CPU inputs the data from the PIA. The CPU then uses the inputted number in a calculation to determine if a switch opening has taken place. If so, a branching statement in the program is executed. If not, nothing occurs, and the CPU signals the PIA to input another number from the input port.

The CPU must poll (ask for data) from the PIA often or there will be a time delay between the switch closure and the change in execution of the computer program. In Figure 6.6 the PIA is polled 50 times per second and data are sent from the PIA to the CPU on the data bus 50 times per second. Therefore, a delay of .020 s can occur between the pressing of the switch and the change in the execution of the computer program.

This technique is an example of a software interrupt. There is also a hardware interrupt method. With a hardware interrupt, the

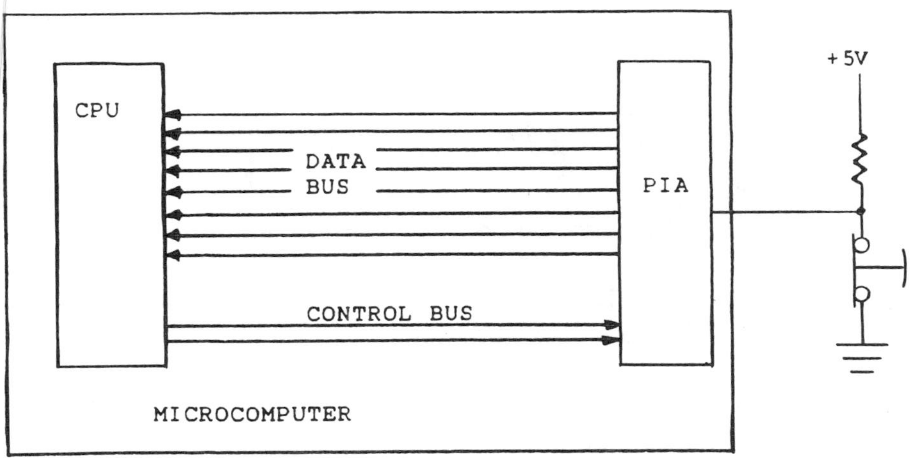

Figure 6.6 Polling to detect a switch closure.

external switch is connected directly to the CPU. A logic high causes the microprocessor to change its execution of a program (starting or stopping execution or branching to another part of the program or to a subroutine). Figure 6.7 shows a 8085 microprocessor CPU with an external pushbutton switch connected to an interrupt pin.

The timing of interrupts and the software requirements of discrete (switch closure) I/O are greatly simplified in programmable logic controllers. Explanation of the theory in this and the other chapters is provided to aid in understanding interfacing.

6.4 DIGITAL INPUT

At times, a machine operator needs to change the value of numbers within the computer program. Analog output values, such as set points for valves, timer set points, and counter limits, are examples of data that must be entered into memory locations within the microcomputer.

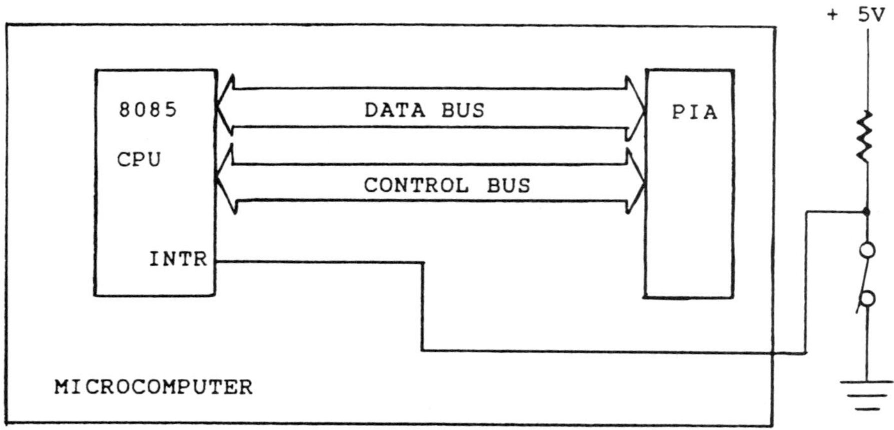

Figure 6.7 Hardware interrupt.

These values can be programmed into the computer when the program is written, or the program can be modified to change the values. These methods are acceptable if it is not necessary to change the values often. If the operator needs the capability of changing a set point during the duty cycle, a more convenient method is needed. BCD switches provide this method.

A BCD (binary-coded decimal) switch is a means of entering a binary number into a microcomputer. The most common is a thumbwheel BCD switch. It consists of three or four plastic wheels that have electrical contacts. The contacts touch combinations of contacts on a printed circuit board within the switch. The electrical output of a BCD thumbwheel switch is a series of binary numbers, as shown in Figure 6.8.

Each of the digits, the numbers 9, 6, and 1, are coded onto a 4 bit binary number by the thumbwheel switch. The code in this example is referred to as 8-4-2-1. Note that the 12 binary digits represent three separate decimal numbers and that this is not the 12 bit binary number for the value 961.

Decimal number, 961
BCD, 1001 0110 0001
Binary, 001111000001

The parallel input port of the microcomputer can accept BCD input, such as the three groups of 4 bits representing the number

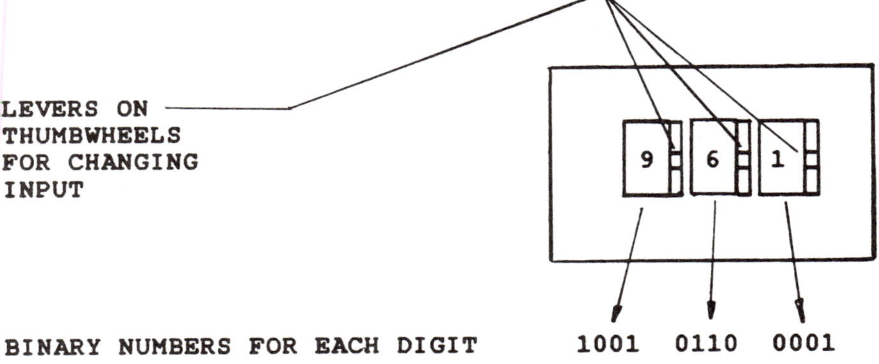

Figure 6.8 BCD input switch.

961. However, the BCD number must be converted into a true binary number before it can be used. This can be done by a calculation in the CPU.

6.5 SERIAL INPUT

This was explained in Chapter 5.

6.6 FREQUENCY INPUT

Speed sensors, such as magnetic pickups and Hall effect sensors, are commonly used to monitor the rotational speed of shafts. The output from these sensors is usually a train of pulses. The frequency of the pulses is proportional to the rotational speed of the shaft (see Chapter 11).

To input a frequency signal into a microcomputer, the signal must be converted into a binary number. Several techniques can be used to accomplish this.

One technique is to convert the frequency signal to an analog voltage and then perform an analog-to-digital conversion (Figure 6.9). This is the technique utilized in a typical frequency-to-voltage converter [2], the Analog Devices 451. It receives a train of pulses from a magnetic pickup and converts the pulses into a series of pulses of constant height and duration. The pulses are then integrated by an operational amplifier circuit. The pulses from the magnetic pickup can be integrated directly, but the result is inaccurate. Red Lion Controls [3] markets a pulse rate to analog converter that operates on a similar principle, the PRA 1.

If the pulses from the speed sensor are TTL compatible (true logic pulses), a digital counting circuit can be used to obtain a binary number. In Figure 6.10 a 4 bit binary counter, a Signetics 7493, is used to count the pulses from an incremental encoder. The pulses are TTL-compatible logic pulses. The output from the left 7493 is used as a divide-by-16 counter. This output is sent to the input of another 7493 counter to produce a divide-by-256 counter. The 4 bit output from the second counter is sent to a parallel port.

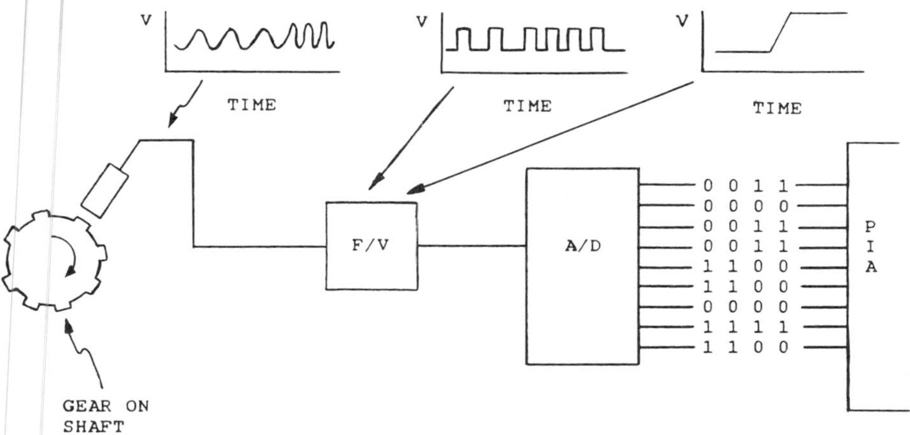

Figure 6.9 Frequency-to-binary conversion using analog voltage and A/D.

The computer then polls the port and the second 7493, at timed intervals, and does a calculation to determine the shaft speed.

6.7 ANALOG INPUT

The input of analog data into a microcomputer, such as 0–5 V or 4–20 mA, requires a conversion. The analog voltage is converted into parallel digital data. Many analog-to-digital converters on the market perform this function. Three popular types are the integrating, successive approximation, and parallel.

The integrating A/D converter (Figure 6.11) uses the timing capability of a digital circuit and the comparison capability of a comparator operational amplifier circuit to convert an analog voltage to a digital number. It does this by starting an analog voltage ramping circuit that increases the voltage going to the comparator at a constant rate. At the same time as the ramping circuit begins, a counter in the digital logic circuit of the converter begins a digital count. As the ramping circuit voltage reaches the level of the test voltage, the comparator operational amplifier sends a

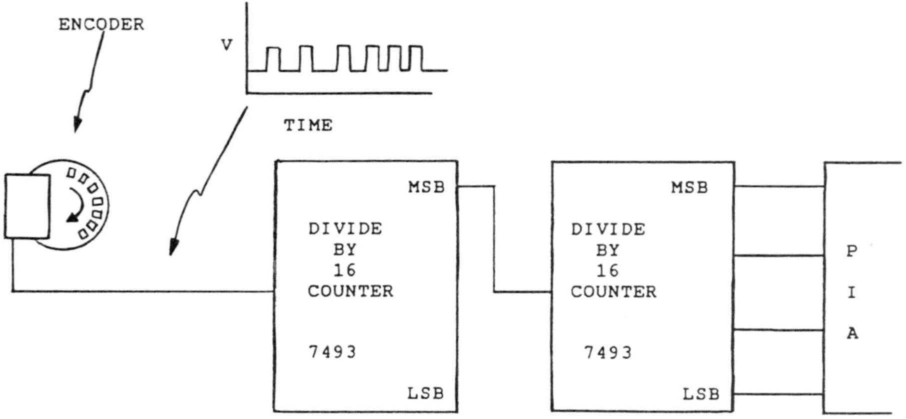

Figure 6.10 Frequency-to-binary conversion using counters.

signal to the counting circuit to stop the count. The resulting digital number is an accurate conversion of the test voltage.

Integrating A/D converters have the highest resolution of the common A/D converters and are the slowest—several milliseconds for a conversion. A variation of this type, which is more accurate, is the dual-slope integrating A/D converter.

The successive approximation A/D converter is faster than the integrating type, less than 100μ. It converts an analog voltage with a guessing-game approach using a D/A converter and a comparator (Figure 6.12).

It begins by setting all the bits entering the D/A converter high, except the most significant bit. This sends a binary number to the D/A that is one-half the maximum number possible. In an 8 bit system this midpoint number is 128. The maximum number is 255.

A voltage is output from the D/A to the comparator. If the voltage is higher than the unknown voltage, the controller tries a smaller number in the D/A. This next number is halfway between 128 and 0, the number 64. The same process is repeated, and the converter converges on the right answer in a few steps.

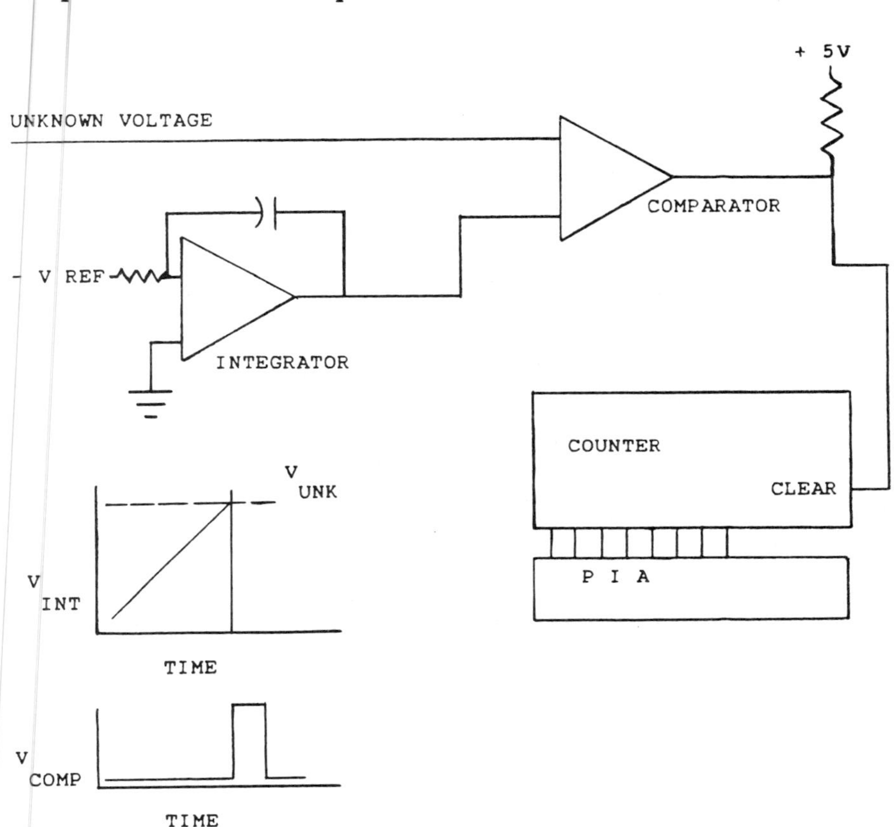

Figure 6.11 Integrating an A/D converter.

A sample and hold amplifier is required with successive approximation A/D converters to hold the analog input value during the conversion. These are usually on the D/A chip, but if an external sample and hold amplifier is needed, an Analog Devices AD582 is an example of this component.

The parallel A/D converter utilizes a simple circuit consisting of a number of series resistors and comparators (Figure 6.13). Starting at V_{ref}, the voltage drop across each resistor results in a lower reference voltage applied to successive comparators. If the analog input voltage is greater than the V_{ref} input, the output from

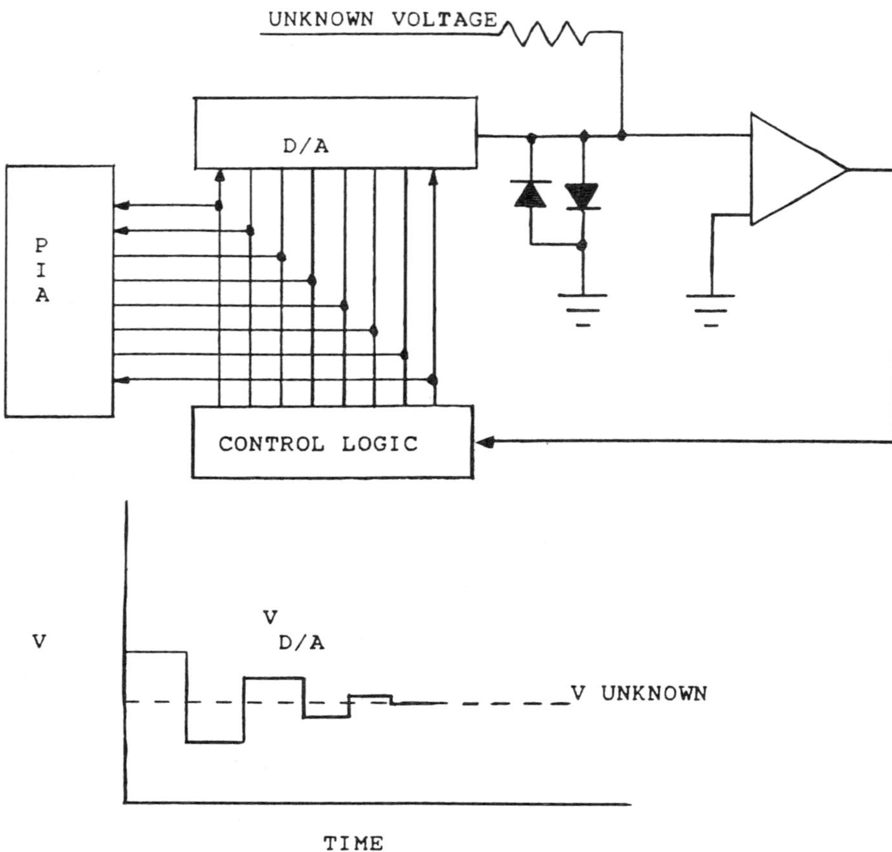

Figure 6.12 A/D conversion by successive approximation.

the comparator is low. The outputs must be decoded to result in a binary number. This is the fastest of the converter methods, 10 ns for a conversion. It has the disadvantage of requiring a large number of comparators, 2 N - 1). In other words an 8 bit converter requires 255 comparators [4].

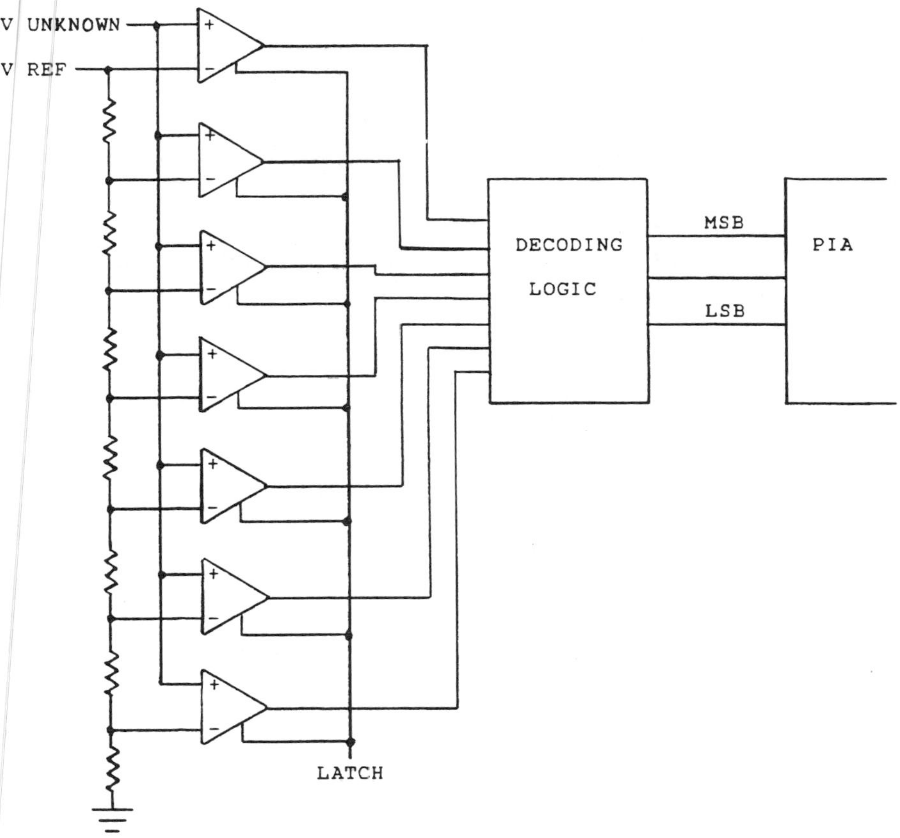

Figure 6.13 Parallel A/D conversion.

REFERENCES

1. Horowitz, Paul, and Hill, Winfield, *The Art of Electronics*, Cambridge University Press, 1980.
2. Analog Devices, Inc., P. O. Box 280, Norwood, MA 02062.
3. Red Lion Controls, Inc., Willow Springs Circle, RD 5, York, PA 17402.
4. Prensky, Sol D., and Seidman, Arthur H., *Linear Integrated Circuits and Applications*, Prentice-Hall, Englewood Cliffs, NJ, 1981.

BIBLIOGRAPHY

Meiksin, Z. H., and Thackray, Phillip C., *Electronic Design with Off-the-Shelf Integrated Circuits*, Prentice-Hall, Englewood Cliffs, NJ, 1984.

7

Electrohydraulic Amplifiers

7.1 INTRODUCTION

To control a hydraulic system electronically, the flow of electric current must be transformed into hydraulic flow or pressure. In electrohydraulic proportional valves, this is accomplished by a solenoid coil. Current flow through the coil causes a force on an armature, which either positions a metering spool to control flow (Figure 7.1) or pushes on a poppet to control pressure (Figure 7.2).

Electrohydraulic servovalves also utilize a coil, but it is arranged to provide torque, which acts upon a jet flapper valve (Figure 7.3). The pilot pressure is then used to position the main metering spool of the valve.

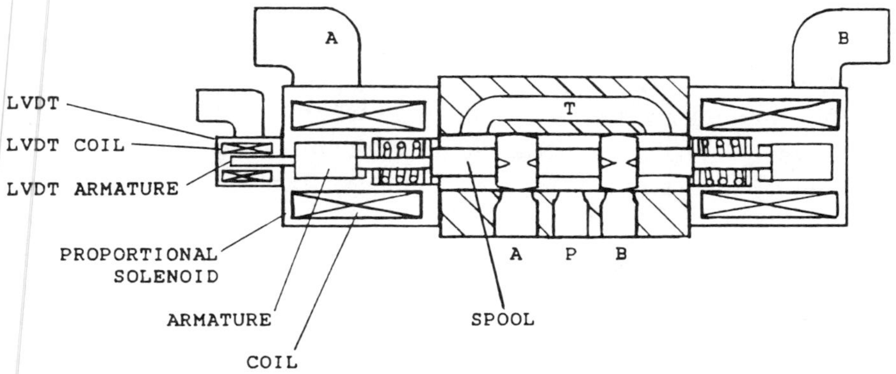

Figure 7.1 Proportional valve with bidirectional solenoid and LVDT.

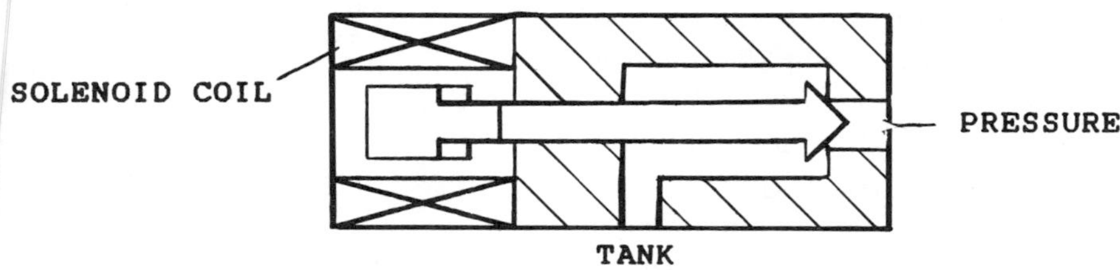

Figure 7.2 A solenoid coil applying force to a poppet to control pressure.

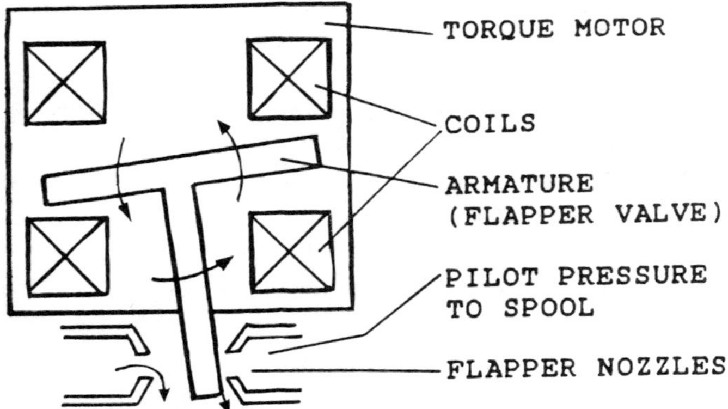

Figure 7.3 Servovalve torque motor with flapper valves.

The electronic component that provides electric current to the solenoid is an electrohydraulic amplifier. These amplifiers are usually supplied by the hydraulic valve manufacturer and are matched closely to the performance of the particular type of valve (Figure 7.4).

7.2 PACKAGING

These amplifiers are usually packaged on a single circuit board. Screw terminals, edge connectors, or DIN pin connectors are used depending upon the manufacturer's preference.

Electronically, these cards consist of analog circuits, that is, linear integrated circuit chips with operational amplifiers and a power transistor. Some amplifiers contain a dc power supply, but others require an external supply.

7.3 FUNCTIONS

These components perform a variety of functions: Line current (115 VAC, 60 Hz) must be rectified and stepped down to voltages compatible with the IC chips and solenoids. Some amplifier cards

Figure 7.4 Electrohydraulic amplifier card. (Courtesy of Racine Fluid Power, Bosch Group, Racine, WI.)

contain a power supply on the card for this, but other systems require a separate power supply.

A command signal is received from a programmable controller, or potentiometer, and amplified into a controlled current for the solenoid coils. Typically, the current is compensated for changes in coil resistance. The current to the coils (and therefore the solenoid force) does not change as the resistance increases because of temperature.

The current to the coils is intermittent. It is either a pulse-width-modulated direct current (Figure 7.5) or a dc with a dither signal superimposed on it (Figure 7.6). The dither signal is required in an electrohydraulic valve to keep the poppet or spool modulating, or moving. This increases the accuracy of the valve by reducing stiction and hysteresis.

Some amplifier cards supply both solenoids of a bidirectional proportional valve (Figure 7.1). These cards receive a dc command

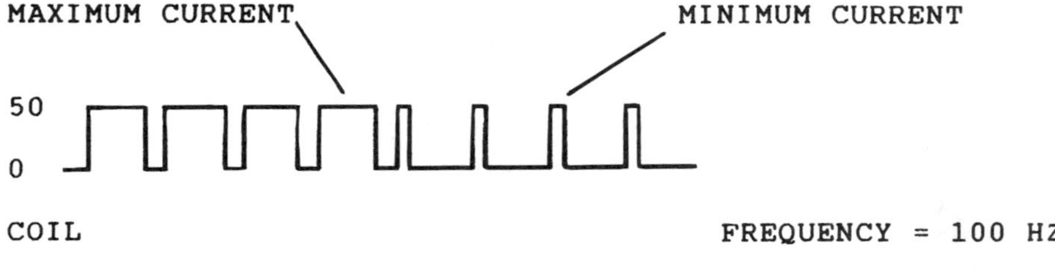

Figure 7.5 Pulse width modulated DC.

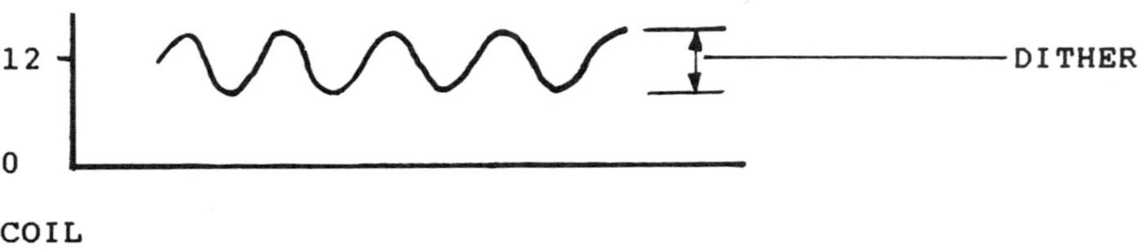

Figure 7.6 Direct current with dither.

voltage in a range +10 to −10 V, for example, and energize either solenoid based upon the polarity of the signal. A flow versus command voltage curve for this valve is shown in Figure 7.7. Note the discontinuity about the center position. This occurs in proportional valves because of the lap of the spool and the lands about the center position of the valve. The lap is unavoidable owing to the economics of manufacturing valves in the proportional price range. The amplifier cards can be made to eliminate the effect of this with a deadband eliminator circuit. The flow-voltage curve then appears as shown by the dashed line in Figure 7.7. The deadband, or lap, in the center position remains, but the amplifier card shifts the spool rapidly through this space by providing high-current gain about the center position.

Another amplifier function is closing inner control loops, such as valve spool position. Many proportional valves and pumps utilize a LVDT (Figure 7.1) to sense spool position or, in the case of a pump, cam ring or swash plate position. The amplifier card supplies the LVDT with ac current and receives a signal voltage proportional to the position of the spool. The position signal is then compared with the command voltage in a summing circuit. An error in the spool position is amplified in the summing circuit and added to the command voltage to correct the spool position.

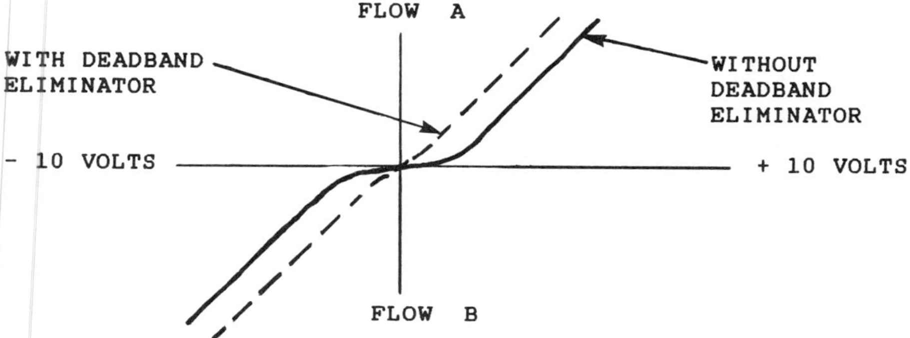

Figure 7.7 Flow versus command voltage bidirectional proportional valve.

Some amplifiers have circuitry to close an outer control loop, such as the position loop in Figure 7.8. The amplifier compares the feedback voltage from a linear potentiometer with the voltage from the command potentiometer. It then amplifies the error and changes the current to the valve to correct the position of the cylinder rod.

Another useful function of the amplifier cards is providing ramping—acceleration and deceleration control. The amplifier, after receiving a step change in the command signal, changes the current to the coil of a valve at a preset rate (Figure 7.9). In a hydraulic system, this results in a controlled rate of change of the flow rate to a cylinder and therefore controlled acceleration of the cylinder rod.

7.4 ADJUSTMENTS

Amplifier cards have adjustments to match the available command signal to the desired value of valve or pump output. Some adjustment is always necessary in proportional valves to compensate for manufacturing tolerances. Figure 7.10 shows the effect of bias and gain adjustments.

7.5 INTERFACING CONSIDERATIONS

Most amplifier cards do not possess digital communication capability. Some manufacturers provide accessory cards for digital I/O, and others rely upon the user to supply the serial-to-parallel and D/A conversion components.

The analog command voltage circuit must also be supplied by the user. Several methods are used to accomplish this. Some methods employ a microcomputer and some do not.

The most simple command voltage method is with a potentiometer in a voltage divider circuit (Figure 7.11). The position of the wiper results in a voltage to the command terminal of

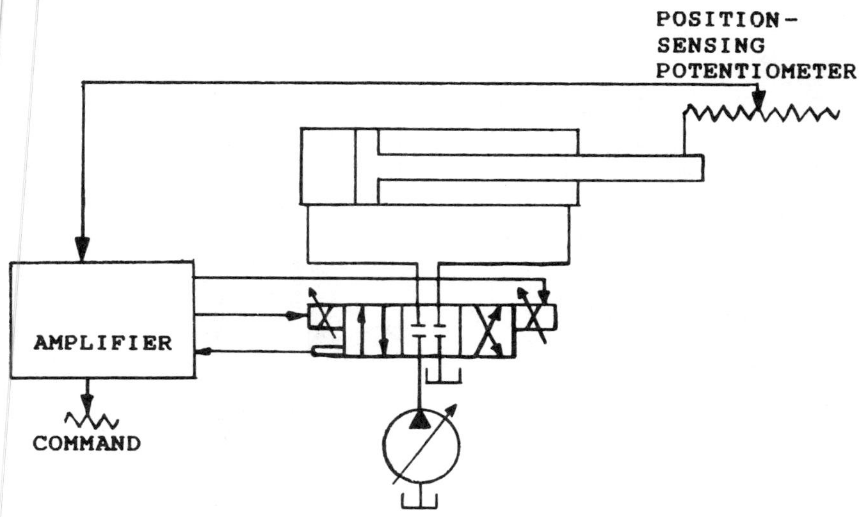

Figure 7.8 Closed-loop control of position.

$$V_{command} = \frac{V_s R_2}{R_1 + R_2}$$

Another method also uses a voltage divider, but a programmable controller is used to sequence between the presets (Figure

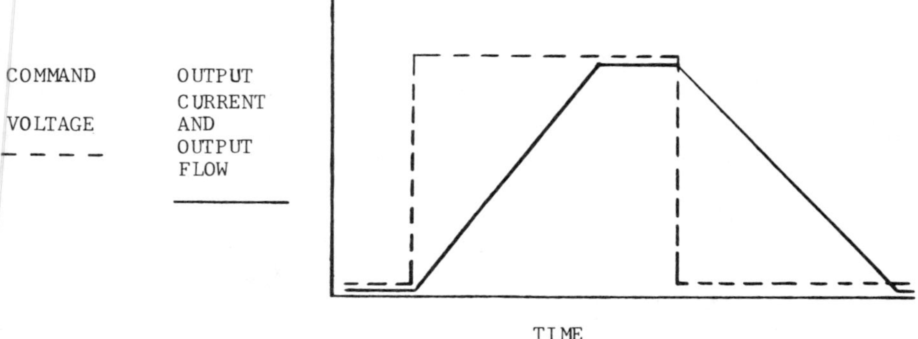

Figure 7.9 Acceleration control with an electrohydraulic amplifier.

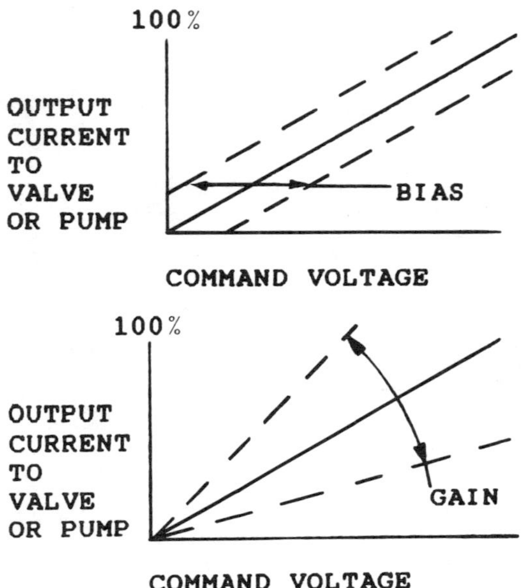

Figure 7.10 The effect of bias and gain adjustment on amplifier output.

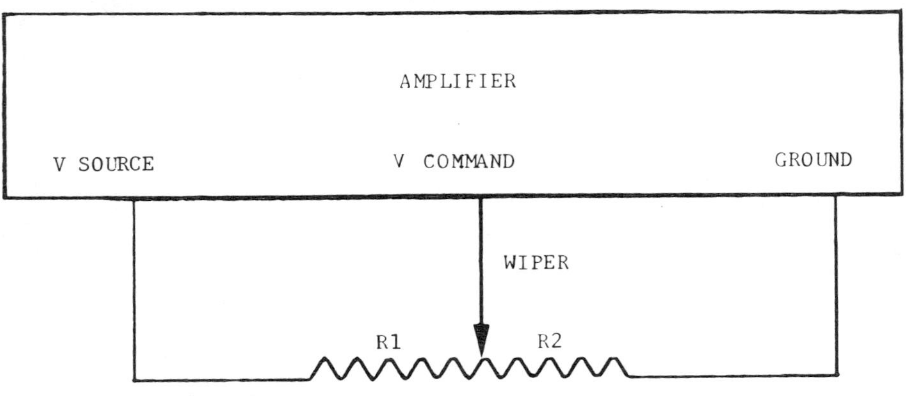

Figure 7.11 Voltage divider command circuit.

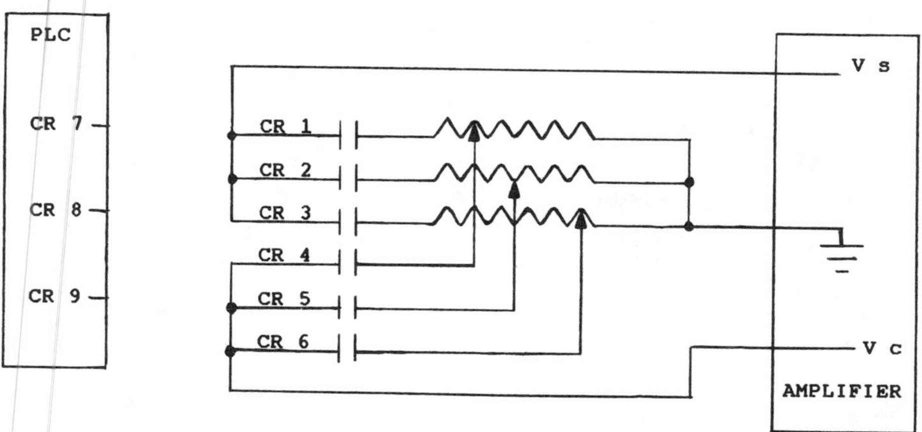

Figure 7.12 PLC sequencing of three presets.

7.12). Three command potentiometers are selected by discrete outputs from a PLC. The PLC outputs CR 7, 8, and 9 activate the control relays CR 1 and 6, CR 2 and 5, and CR 4 and 5, respectively.

The relays CR 1–6 can be eliminated if outputs CR 7–9 are double throw contact closure outputs. Some manufacturers supply control relays and command potentiometers on the amplifier cards, and some mount these on an additional card.

Programmable controllers with analog output are also used to supply the command signal. The controller output is typically 0–5 V, or a 4–20 mA current loop. A 4–20 mA current can be converted into a command voltage of 1–5 V by passing the current through a 250 Ω resistor.

Dry contact relays, such as reed relays or mercury-wetted relays, are required for reliable switching in command voltage circuits. The reason for this is that the low current in these circuits is insufficient to burn off oxidation or dirt on the contacts. The low voltage is insufficient to overcome the resistance of an oxide layer on the contacts. Reed relays are sealed from the environment, preventing oxidation and contamination of the contacts.

7.6 SUMMARY

Amplifier cards are analog circuit cards that amplify a command voltage (0–5 V typically) into a controlled current (0–1 A typically) for the solenoid coils. Adjustments are provided to compensate for manufacturing tolerances and for user convenience. Simple functions, such as acceleration and deceleration, can be added to the cards.

8

Electrohydraulic Valves

8.1 INTRODUCTION

Hydraulic valves are used to direct and restrict the flow of fluid in a hydraulic circuit. Electrohydraulic valves are hydraulic valves that perform these tasks according to electrical command signals received from electronic controllers. Various types of electrohydraulic valves perform similar functions at various levels of speed and accuracy. In this chapter, various types of electrohydraulic valves are discussed from the simplest proportional to the most sophisticated servovalves.

8.2 PROPORTIONAL VALVES

Proportional valves came about by the attempts of many to build a low-cost alternative to a servovalve. Industrial directional valves were modified and given electronic controllers. To better understand the operating principles of a proportional valve, a design discussion is presented here in which a four-way directional valve is redesigned into a proportional valve.

In Figure 8.1 a common industrial four-way directional valve is shown. The direction of the fluid flow through the valve is con-

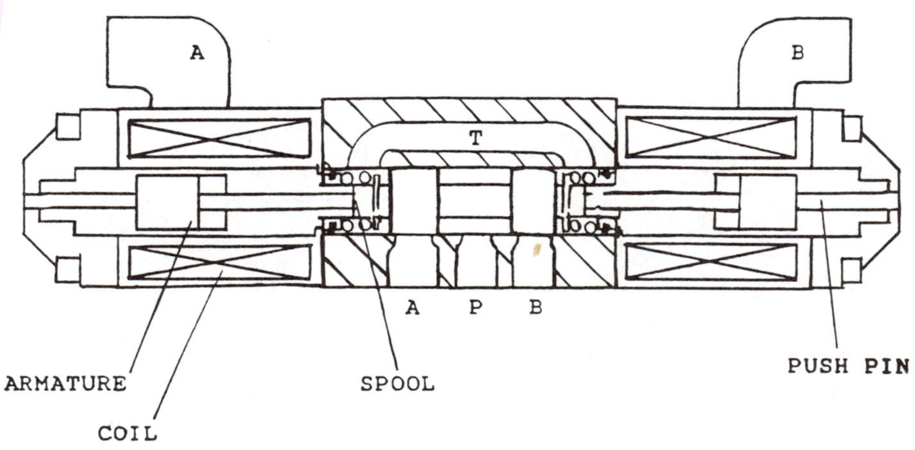

Figure 8.1 Industrial four-way directional control valve.

trolled by activating a solenoid on either end of the valve. The valve has three positions and the flow is off, or maximum, in one direction or the other. The valve does not meter the flow (50% flow, for example) as a flow control valve would.

To convert this valve to a proportional valve, it must be redesigned to achieve the following:

1. Enable the spool to be positioned in an infinite number of positions between one end and the other.
2. The flow rate through the valve must be proportional to an electrical command signal.

The first step of this redesign is to change the solenoid to a proportional design. The solenoid of a directional valve has an air gap through which magnetic flux lines pass. In a two-position valve the force exerted by the solenoid is proportional to the air gap (Figure 8.2). In a proportional solenoid the air gap is changed so the force exerted by the armature is constant and does not change with spool position. The constant force-air gap relationship allows the designer to regulate the force of the armature with the current through the coil. The air gap distance has no effect that would change the current-force relationship.

Metering notches are added to meter the flow at valve positions near the opening and closing positions. Figure 8.3 shows a force-controlled proportional valve. The force of the solenoid is

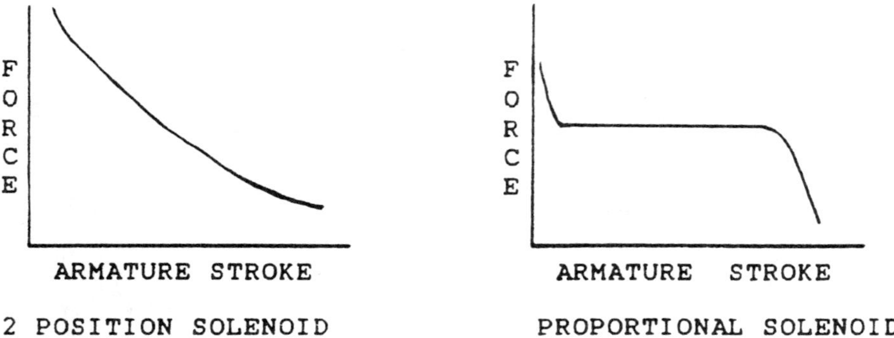

Figure 8.2 Comparison of force curves of valve solenoids.

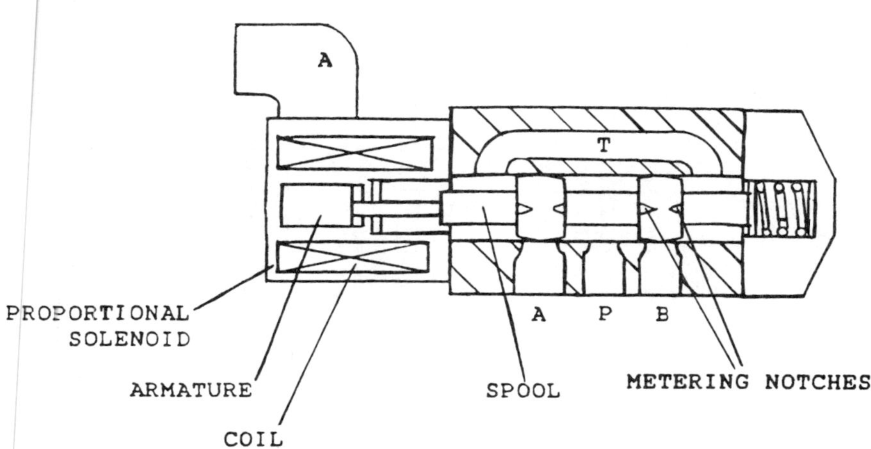

Figure 8.3 Proportional valve with force-controlled solenoid.

proportional to coil current, and the spool position is proportional to the force because of the opposing spring on the other end of the spool.

This valve has an accuracy problem caused by flow forces within the valve. Because the spool position is determined by force, flow forces can change the force balance between the solenoid and the spring, moving the spool to a different position.

This problem is solved by a proportional valve in which the spool is positioned by a positioning means rather than a force balance. This is accomplished in the proportional valve shown in Figure 8.4.

An electrical position sensor has been added to the proportional valve to monitor spool position. A linear voltage displacement transducer (LVDT) sends a feedback signal back to the amplifier supplying the current to the solenoid coil. The spool position signal is then compared with the command signal to determine if the spool is in the commanded position. If it is not, the amplifier changes the current level to the coil, which results in spool movement to the commanded position.

This is closed-loop spool position control with electrical feedback. With this, the accuracy of flow control of the valve is im-

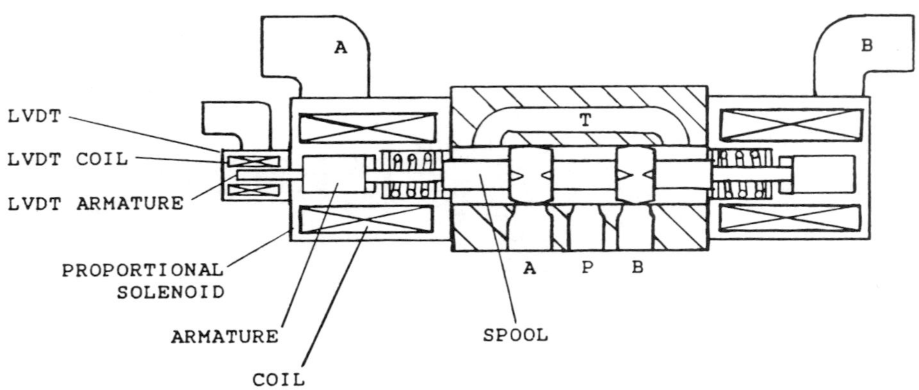

Figure 8.4 Proportional valve with position-controlled solenoid and LVDT.

proved by elimination of spool position errors caused by the rate of the control spring, force variations on the spool from flow forces, and variation in the linearity of the solenoid force. Proportional relief, pressure reducing, and flow control valves also benefit from position-controlled spools.

However, even with these improvements to control spool position accuracy, in most proportional valves a great source of flow control inaccuracy remains; the cast lands in the valve body. One of the reasons for the low cost of proportional valves is common manufacturing techniques with directional valves. The lands (or the port openings) of a directional valve are cast. In most proportional valves these are also cast. Tolerances of position in castings are typically no better than $\pm.020$ inches. It is impractical to manufacture a unique spool for every valve, so the designer must live with a $\pm.020$ inch tolerance in the valve timing. The spools of proportional valves have tapered lands that somewhat reduce timing errors, but this is a compromise solution.

Some manufacturers of proportional valves add a sleeve to the valve between the spool and the body (Figure 8.5). The machined ports of the sleeve control valve timing more accurately than cast lands. This technique adds cost, and this must be weighed against the need for the improved performance. Additional comments

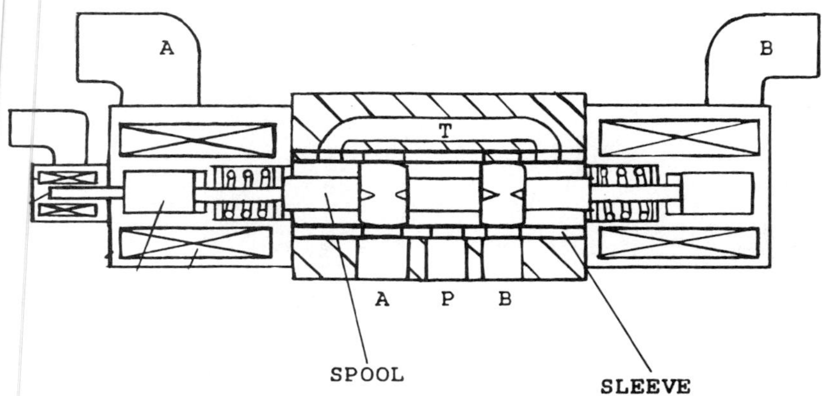

Figure 8.5 Proportional valve with sleeved ports.

regarding cost and performance appear in the servovalve discussion.

8.3 DEADBAND

In Figure 8.6, a flow versus command voltage (or flow versus spool position) curve is presented. Note that there is a discontinuity or interruption in the curve at the centered position of the spool. This is referred to as deadband. This is because of the overlap in the center position of the proportional valve.

The deadband is not present if the valve spool has zero lap. In a zero lap valve, spool movement in either direction from the center position results in flow through the valve. The curve approaches the ideal of Figure 8.7. Owing to manufacturing tolerances in directional valves and proportional valves, it is necessary to produce valves with overlap. Flow versus command voltage curves for overlapped, zerolapped, and underlapped valves are presented in Figure 8.7.

Center lap conditions are important in reversing and closed-loop position applications. In a reversing application an underlapped condition may provide superior smoothness of opera-

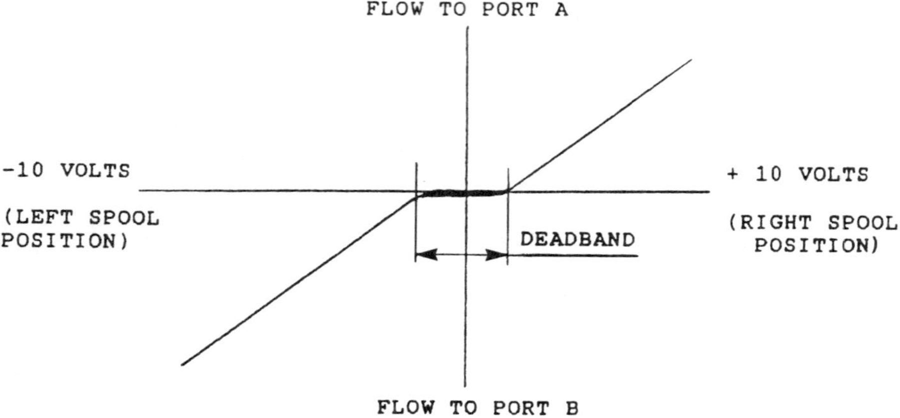

Figure 8.6 Flow versus command signal and spool position in an electrohydraulic proportional valve.

tion during reversal compared with an overlapped spool, which can cause abrupt flow changes or surges.

The center lap condition is a critical difference between low-cost proportional valves and servovalves. The zero lap condition is expensive to achieve. Manufacturers of higher cost proportional valves achieve a zero lap by accurately machining ports into a sleeve that surrounds the spool. The sleeve performs the timing functions instead of the cast lands in lower cost proportional valves.

A zero lap condition can be simulated electronically by a circuit called a deadband eliminator (Chapter 7). The overlap remains but the electrohydraulic amplifier card of the valve eliminates the command voltage levels of the deadband. A spool commanded to the deadband is commanded by the amplifier to a position on one side or the other at the point at which flow is about to begin.

Larger valves require higher forces to position the spools. The size of the solenoid becomes impractical above a certain level, and pilot-operated valves are necessary. A proportional pilot valve is similar in design to a single proportional valve. The main valve of a pilot-operated proportional valve may also have a LVDT for

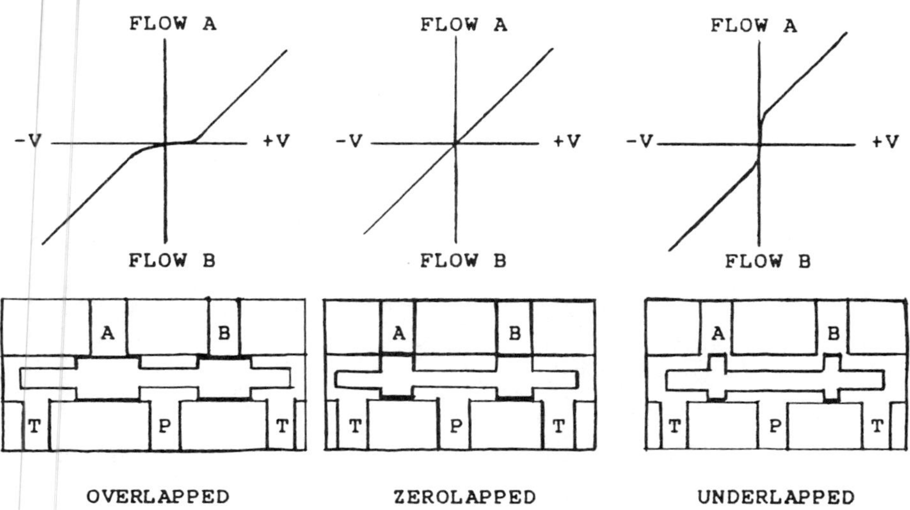

Figure 8.7 Flow curves for various center lap conditions.

spool position. In this case the amplifier closes two inner loops (Figure 8.8).

8.4 PROPORTIONAL PRESSURE CONTROL VALVES

A simple pressure control valve can be constructed by replacing the screw and spring of a mechanically adjusted relief valve with a proportional solenoid assembly (Figure 8.9). The force from the solenoid armature exerts a force on a poppet instead of the screw and spring. This method is limited in the flow rate that can be handled, so it is commonly used in pilot sections of two-stage electrohydraulic pressure or flow control valves.

A method that uses a positioning technique to control pressure is shown in Figure 8.10. The armature of the solenoid applies force to a spring seat, compressing a spring. As the spring seat moves, a LVDT senses its position and feeds it back to the electrohydraulic amplifier. The precise positioning of the spring seat results in a precisely controlled force on the poppet.

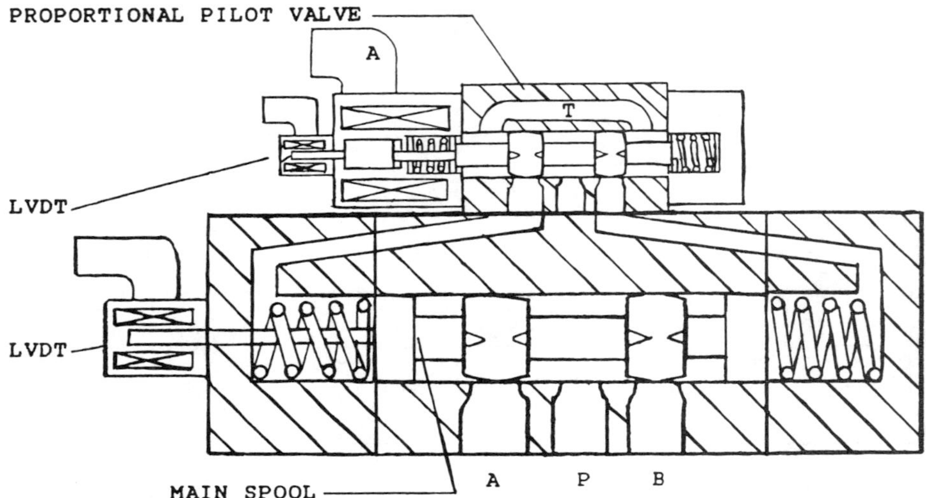

Figure 8.8 Pilot-operated proportional valve.

8.5 SERVOVALVES

Electrohydraulic servovalves have been in use since the 1950s. These were the traditional electrohydraulic valves used in industrial, aerospace, and other applications.

Historically, there have been distinct differences between servovalves and proportional valves. Servovalves were always thought of as high-performance valves in speed and accuracy, and proportional valves were thought of as lower performance, lower cost alternatives to servovalves.

In the 1980s the definitions "servovalve" and "proportional valve" and the distinctions between the two have become vague and overlapping. Lower cost servovalves and high-performance proportional valves are coming closer together in performance and cost. A comparison of servovalve and proportional valve performance is presented in Table 8.1. These values are approximate, and designers should do their own surveys of the market and consider both servovalves and proportional valves in any application that does not demand "extreme" speed or accuracy.

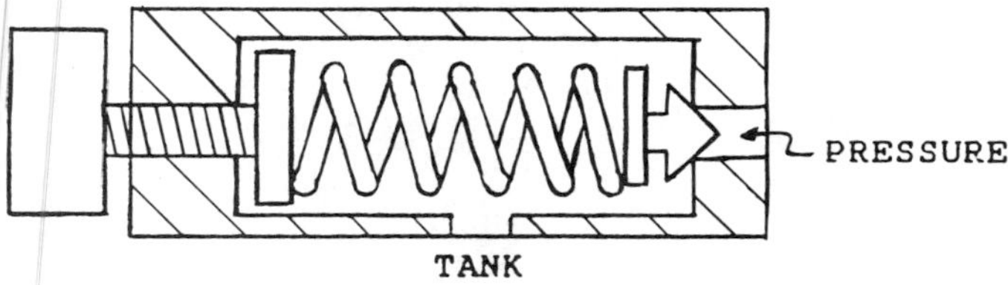

MECHANICAL ADJUSTMENT

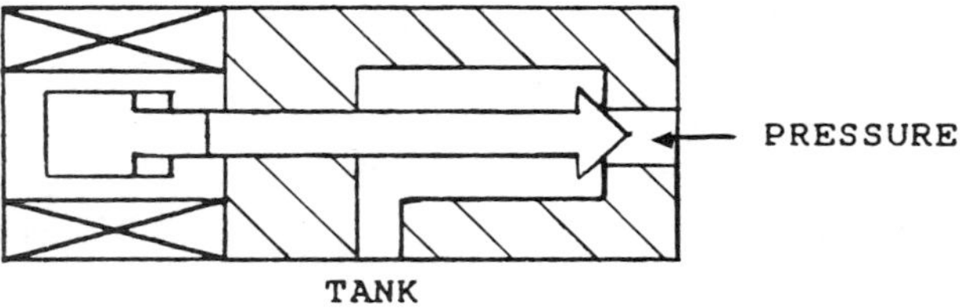

ELECTRICAL ADJUSTMENT

Figure 8.9 Comparison of hydraulic and proportional electrohydraulic poppet valves.

A typical servovalve is shown in Figure 8.11. The flow of current through a coil results in a torque in the torque motor that moves a flapper toward a nozzle. The flapper position between the nozzles is proportional to the current through the coil. There are nozzles on either side of the flapper in a bidirectional valve. There is a constant flow of fluid through the nozzles, impacting the

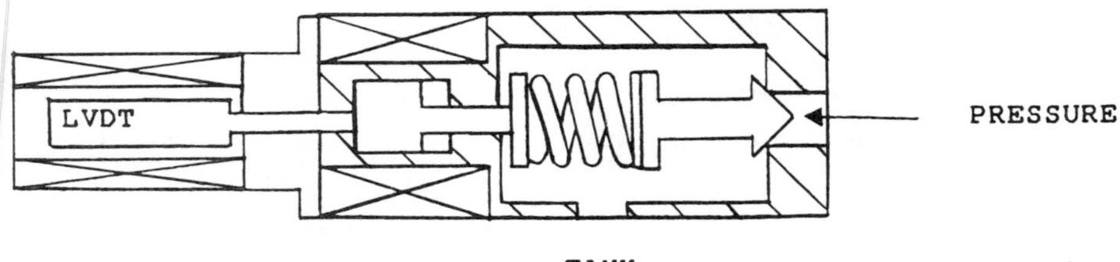

Figure 8.10 Pressure control by a position-controlled solenoid.

Table 8.1 Comparison of Proportional Valves and Servovalves

Specification	Servovalve	Proportional valve
Accuracy (linearity + hysteresis + repeatability) (%)	1	5–15
Speed of response, 0–100% flow (ms)	10	50
Frequency response (Hz at −3 dB)	50–400	5–15
Deadband (% about center)	0–2	>3
Filtration required (μm)	4	10
Most applications open or closed loop?	Closed outer	Closed inner Open outer

flapper. As the flapper moves closer to one nozzle, the pressure in the passage supplying that nozzle rises. The pressure in the other nozzle passage decreases. The imbalance of pressure on the main spool moves the main spool. The nozzles are mounted on the main spool and move with the spool. The spool moves until the nozzles are equal distances from the flapper. The servo is now at a null position, maintaining a flow rate proportional to the coil current.

Other servovalve designs are also in use. Nearly all servovalves share the following characteristics:

Torque motors instead of proportional solenoids
Mechanical feedback from the control spool back to the torque motor
Higher pilot pressure required for operation (1000 psi)
Precise flow metering about the center position of the valve (zero lap)

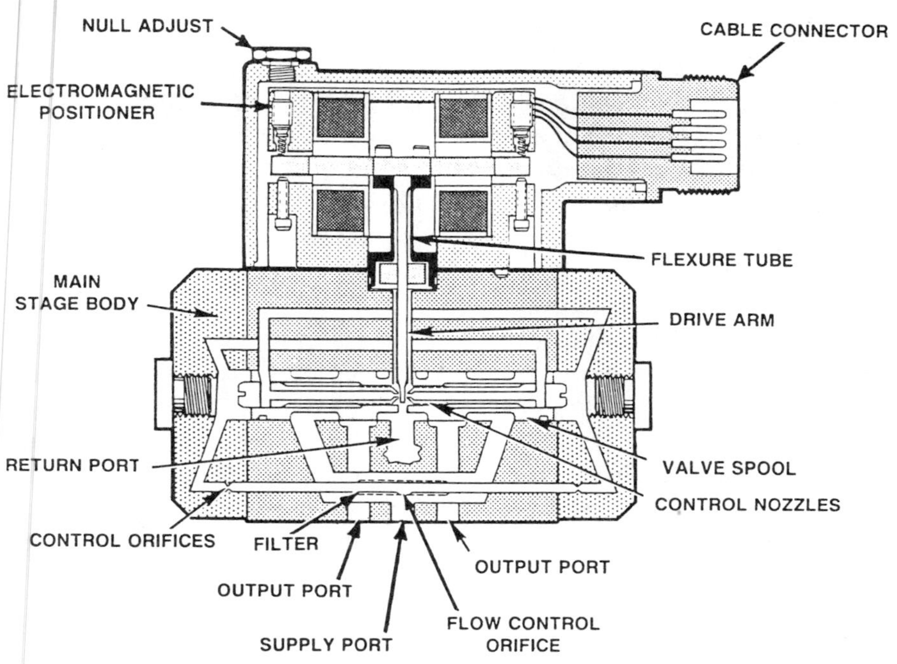

Figure 8.11 Servovalve. (Courtesy of Schenk-Pegasus, Troy, MI.)

8.6 CONTROL LOOPS

In industry, the term "closed loop" is used often, and it is not always clear whether the control loop is within a component (inner loop) or controls a system (outer loop). The distinction is important because of the effect upon system accuracy and complexity.

For clarification, let us discuss closed-loop control, both inner and outer control loops, and how these loops are closed. A closed-loop system or component is one in which a controlled element sends a feedback signal back to a controller. Closed-loop techniques are used to improve the accuracy of components and systems. In component design, the manufacturing tolerances on components can be relaxed somewhat if a closed-loop technique is used.

A closed inner loop is a control loop within a hydraulic component (Figure 8.4). This could be LVDT feedback of spool position or mechanical feedback of a follower servo. The point is that the loop is closed within the component to enhance the accuracy of the component. A closed inner loop does not enhance system accuracy (other than by simply providing a more accurate component of the system).

A closed outer loop (Figure 8.12) consists of a feedback transducer in a system that feeds back a system condition to the controller. The controller can modify the command signals to various components of system to correct an undesired system condition.

In this system the speed of the hydraulic motor can decrease owing to an increase in the load being driven by the motor. In the closed outer loop system, the transducer senses the lower speed and the amplifier increases the pump flow to correct the motor speed. An inner control loop, such as a LVDT on the cam ring of the pump, senses only the pump displacement and cannot sense or correct a system condition like decreasing motor speed.

Control loops can be closed by microcomputers (refer to chapter 12). Closed inner loops in hydraulic components are seldom closed with a computer. The analog circuitry of a typical electrohydraulic amplifier is well developed for this purpose, and cost effective.

8.7 INTERFACING CONSIDERATIONS

In most microcomputer-controlled systems with proportional valves or servovalves, the computer sends an analog voltage or current to the amplifier of the valve (chapter 7). The common voltages used are 0–5 V, 1–10 V, from −10 to +10 V, and 4–20 mA of current.

Another interfacing method is beginning to find use but is not widely used at this time: a serial (digital) interface. In such systems,

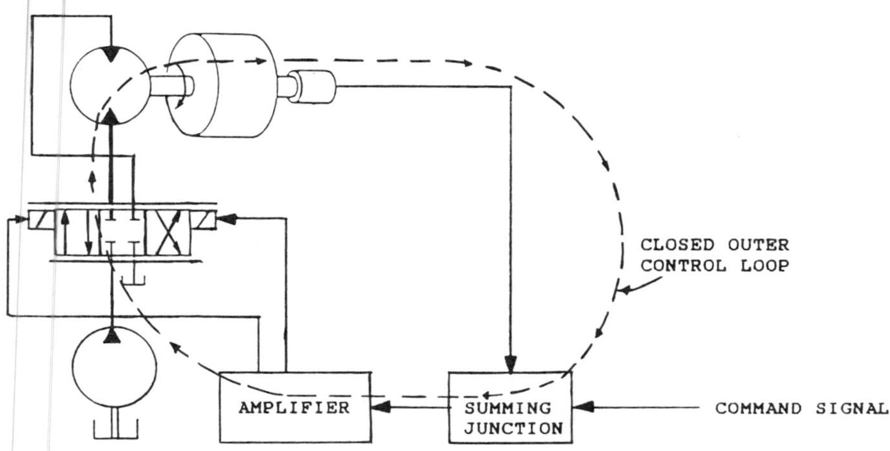

Figure 8.12 Closed "outer" control loop.

the amplifier and digital communication circuitry are mounted at the valve (Figure 8.13).

The microcomputer controlling the system sends a RS 232 or RS 422 serial data signal to the valve through a twisted pair cable. The valve contains a serial-to-analog converter and an electrohydraulic amplifier.

The main advantage of this approach is the reduction in wiring costs. A single two-wire cable of light-gage wire is the only conductor necessary between the controller and the valves. This can be useful in applications in which a central computer controls hundreds of valves mounted hundreds of feet from the controller.

The disadvantages of this approach are many. Troubleshooting of the signal to a valve requires the maintenance person to have a knowledge of digital communication.

Another disadvantage is the cost of the valve. The cost of the digital circuitry (serial or analog) may or may not be significant. Of greater significance is that the valve is no longer a "commodity" or common proportional valve. Valves that receive digital communication are not nearly as common or available as valves that respond to a 4–20 mA current loop. A valve that is harder to get

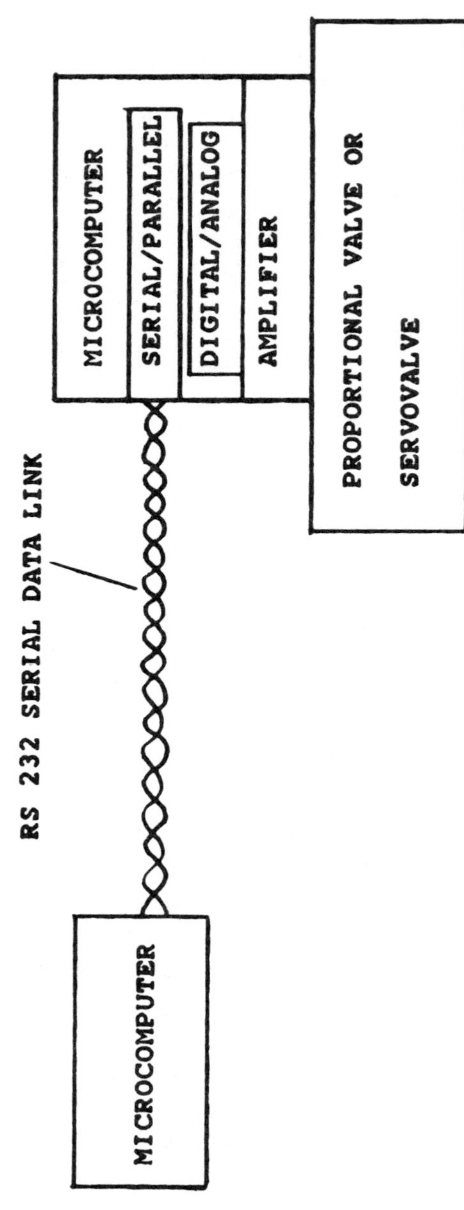

Figure 8.13 Distributed control of a valve.

requires that the user maintain an inventory of spare valves to avoid excessive downtime.

Another disadvantage is the difficulty in supplying a command signal to the valve in the event of controller failure. In an electrohydraulic system using an analog voltage command signal, if a controller failure occurs it is usually simple to operate the valve manually with a potentiometer. In a digitally controlled system, an alternative supplier of the digital command signal is required (another controller or computer).

8.8 STEP MOTOR INTERFACES

Another means of interfacing a computer to a valve utilizes a step motor and a lead screw to position a valve spool. (Figure 8.14). This method has very high resolution in the positioning of the spool, as low as .005%. This is due to the resolution of angular positioning of a step motor. Step motors are available that have a resolution of angular positioning of 7000 positions in one revolution.

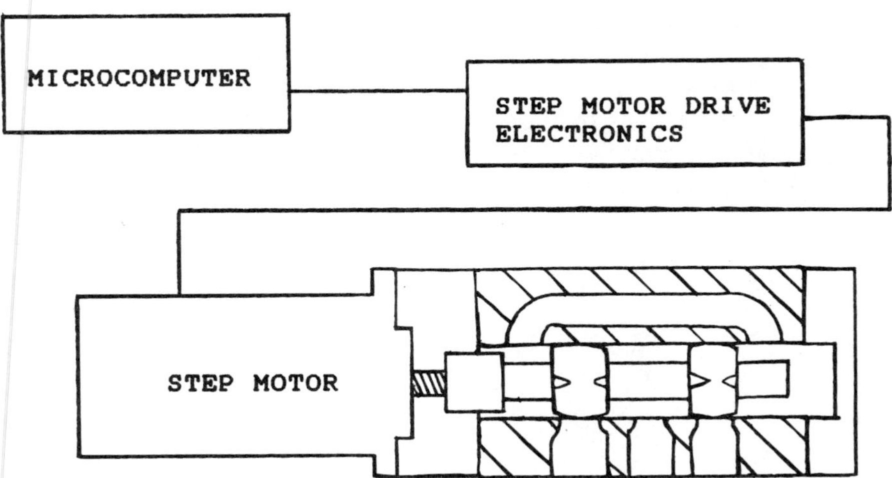

Figure 8.14 Step motor control of a spool valve.

This method has the advantage of eliminating the pilot section of a servovalve and the problems common to this section. Contamination sensitivity is one of these problems, due to the nozzles and flapper valves, which can become clogged.

A disadvantage of a step motor spool positioner is speed of response. The speed of response of this controller is limited by the torque and inertia characteristics of the motor and by the pulse rate from the controller. If the pitch on the lead screw is made coarse enough and the resolution (pulses per revolution) of the step motor is reduced, the speed of response of a step motor spool positioner can approach the speed of a conventional servovalve.

However, control resolution is sacrificed with a coarse lead screw and lower pulse or revolution of the step motor. The control resolution becomes unacceptably large. To increase the resolution, the pitch of the lead screw must be increased along with the pulse or revolution of the stepmotor. The result is once again slow response time. If the designer recognizes this trade-off between speed and resolution, a suitable compromise may be possible for a particular application.

Another advantage of the step motor method is that it is easily interfaced to motion controllers. Motion controllers are specialized microcomputers for positioning applications. If a machine has a motion controller on it to control step motors used for nonhydraulic tasks, one channel of the controller can be used to control the servovalve. This may be a more cost-effective approach than adding analog command circuitry and an electrohydraulic amplifier to the machine.

9

Electrohydraulic Pumps

9.1 INTRODUCTION

Another method of hydraulic system control can be achieved at the power source, the pump. In certain applications energy efficiency and circuit simplification can result from system control at the pump instead of control valves downstream. The purpose of this chapter is to present contemporary methods of electrohydraulic control of variable displacement—industrial hydraulic pumps—and how these methods can be utilized to advantage in a hydraulic system.

9.2 PUMP CONTROL EVOLUTION

The earliest and least sophisticated hydraulic pumps were of the fixed displacement type (Figure 9.1). These pumps had no integral pump control and provided constant flow regardless of pressure. Flow in excess of load requirement was usually diverted through a relief valve.

Figure 9.1 shows a typical fixed displacement pump, a gear pump, in a clamping circuit of a machine tool. The pump, rotating at a constant speed, provides a constant flow into the system. As

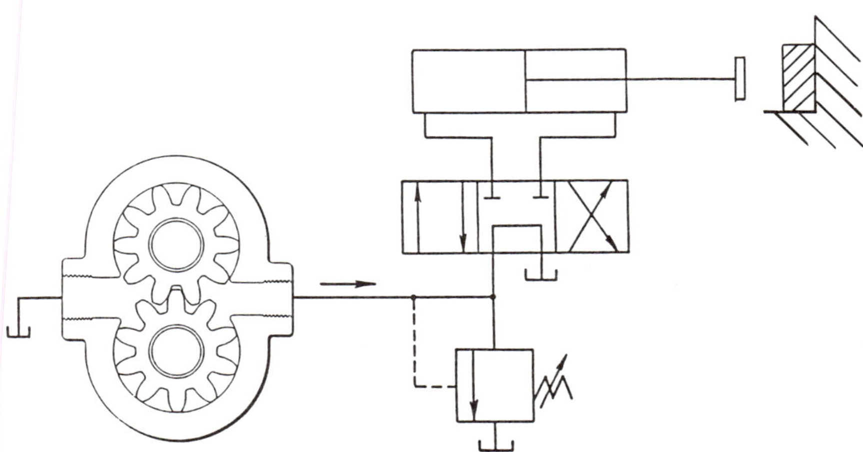

Figure 9.1 Fixed displacement pump.

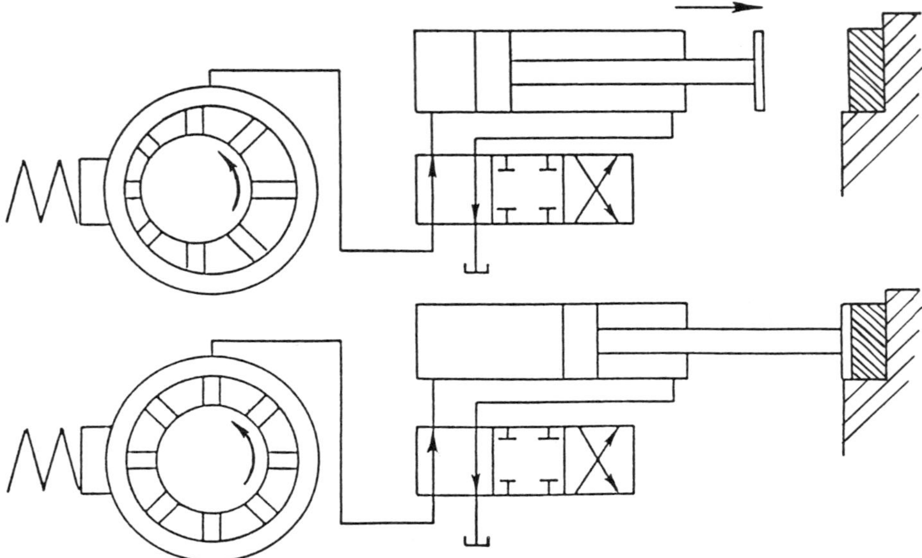

Figure 9.2 Variable displacement pump in full flow (above) and in deadhead, zero flow (below).

the rod of the cylinder moves toward the workpiece, there is little resistance and the relief valve remains closed. As the clamp contacts the workpiece and cannot move farther, the pressure of the system rises quickly. As the setting of the relief valve is reached, the valve opens and diverts the flow back to the reservoir.

This is a classic circuit, simple in concept but wasteful of energy. All the hydraulic power put into the system by the pump is converted by the relief valve into heat that enters the reservoir.

9.3 VARIABLE DISPLACEMENT PUMPS

Variable volume hydraulic pumps were then developed. In the vane pump shown (Figure 9.2), the displacement is variable, controlled by a spring. As system pressure increases due to increasing system resistance (as the rod contacts the workpiece), internal forces acting upon the cam ring push it to a centered position. The pump maintains system pressure with zero output flow. Since no flow occurs, little energy is expended. An additional benefit of this type of pump control is that fewer valves are required in the circuit.

A further refinement of spring-governed pump control was the replacement of the spring with one or two servo pistons and a compensator (Figure 9.3). At pressures below the deadhead setting of the pump, the larger of the two pistons pushes the cam ring to an eccentric (100% flow) position. As the system pressure approaches the deadhead setting, a poppet valve unseats, venting a chamber on one end of the compensator spool. This changes the force balance on the spool, causing it to move to a metering position. Fluid is then metered out of the larger piston chamber, reducing the pressure in the chamber. The smaller piston, which is exposed to system pressure, then overcomes the larger and pushes the cam ring to an almost centered position. The pump maintains pressure without output flow.

Axial piston pumps are controlled in a manner similar to that of the vane pumps. Both spring and servo control are available to control the angle of the swash plate (Figure 9.4).

The advantages of this type of control were control accuracy and stability and flexibility of pump control. Various arrangements of valve and pistons could be attached to the spool valve control to provide additional pump control functions (load sensing and torque limiting, for example). Another advantage was that electrohydraulic valves could be readily attached to the pump for electrohydraulic pump control.

9.4 ELECTROHYDRAULIC PUMPS

Electrohydraulic controls added to a variable volume pump generally perform one of two functions:

1. Regulation of the deadhead pressure set point of the pump
2. Regulation of the displacement of the pump

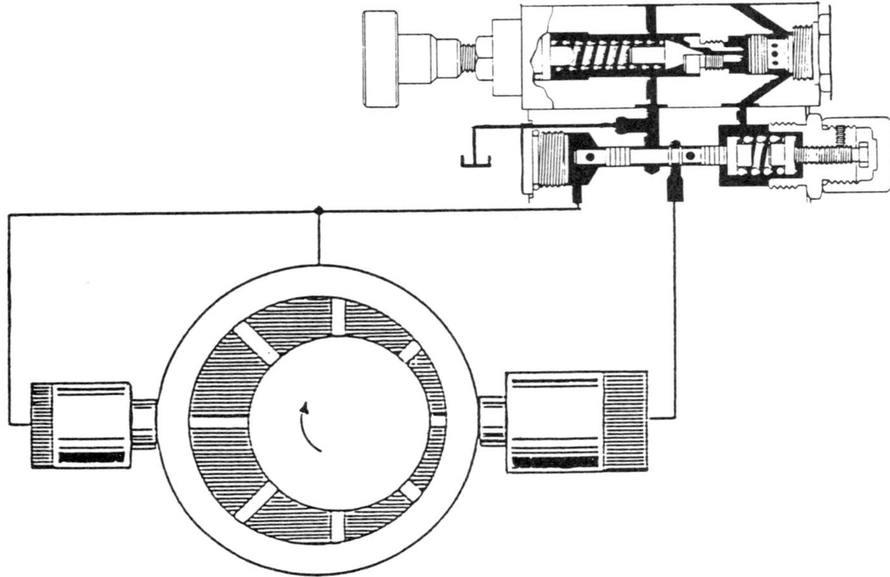

Figure 9.3 Variable displacement pump with two-stage hydraulic control.

9.4.1 Pressure Set Point Control

This is a simple control method by which the mechanical adjustment of the compensator is replaced by an electromechanical adjustment. Figure 9.3 shows a poppet valve of a servo-controlled pump, adjusted mechanically by a screw and spring. Figure 9.5 shows an electrohydraulic alternative, a solenoid armature pushing upon the poppet.

The purpose of this type of control is to regulate pump displacement so that a preset deadhead pressure level is not exceeded in the system. The output flow of the pump is determined by the back pressure of the system and the deadhead pressure set point. As the deadhead pressure is approached, the poppet unseats. This vents the spring chamber of the first stage, changing the force balance on the spool. The cam ring then moves offstroke, reducing pump displacement. As flow is reduced, pump

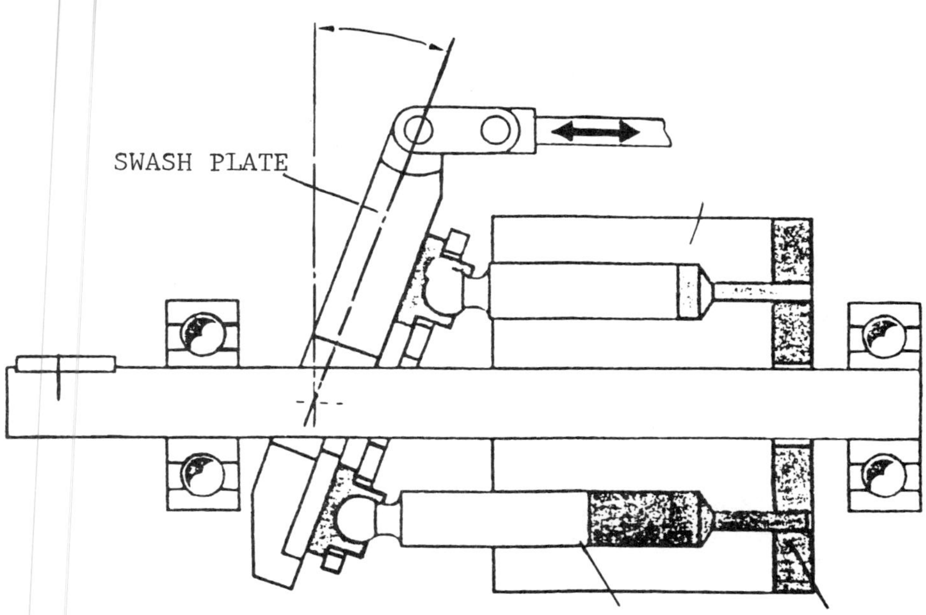

Figure 9.4 Variable displacement axial piston pump.

output pressure decreases. The pump then stabilizes at a ring position that maintains the deadhead pressure in the system. A pump in this modulating condition, maintaining a pressure, is said to be in compensation. The pump acts as a pressure source.

This type of control is commonly used in demand systems. With the addition of electrohydraulic controls, the set point can be changed repeatedly throughout a duty cycle, precisely controlling force within the system, or simply idling the pump at low pressure at times of no demand.

9.4.2 Electrohydraulic Pump Flow Control

This is accomplished by controlling pump displacement. The output flow rate of an axial piston pump can be varied by changing the swash plate angle, as in Figure 9.4. In a variable volume vane pump, the displacement is varied by changing the eccentricity of

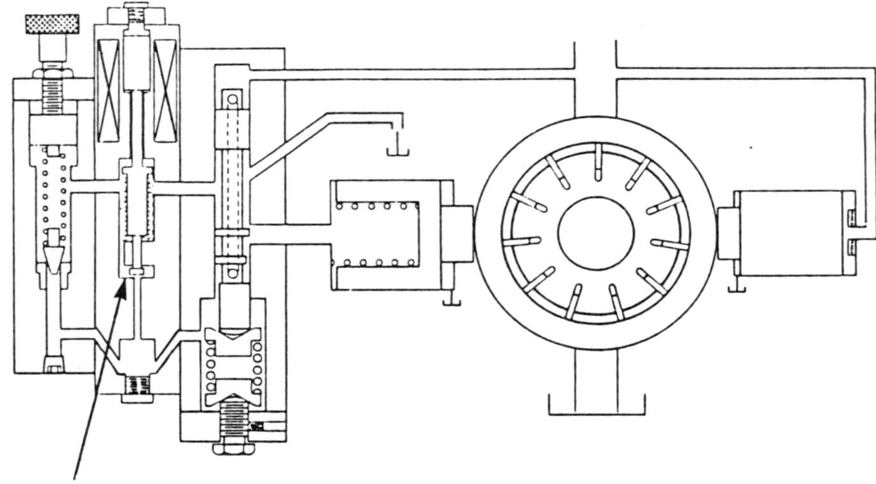

Figure 9.5 Proportional solenoid controlling deadhead pressure. The force output of the solenoid, which is applied to the poppet, controls the compensator setting (arrow).

the cam ring position (Figure 9.3). There are four commonly used methods for electrohydraulic displacement control:

1. Adjustable mechanical stop
2. Flow feedback using an electronic orifice
3. Displacement control with mechanical feedback
4. Displacement control with electrical feedback

9.5 ELECTROMECHANICAL DISPLACEMENT CONTROL

An old, simple method to position the cam ring or the swash plate mechanically was to add a screw to the maximum volume stop of the pump (Figure 9.6). By attaching an electric motor to the screw, electromechanical displacement control could be accomplished. Addition of an adjustment screw is a convenient, simple method to limit displacement but has limitations in application.

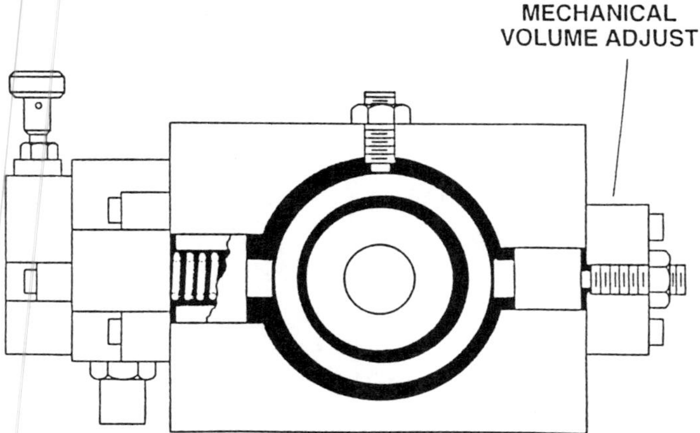

Figure 9.6 Variable displacement pump with mechanical maximum volume stop.

The force levels to turn the screw can get quite high at high pressures. Electric motors used to drive the screw must have high torque capability and therefore large size.

This method is also slow. The large electric motor required has high inertia, which must be overcome. Considering the high forces required to turn the screw (static and dynamic friction) and the rotational inertia of the motor, rapid changes in displacement (0–100% in less than several seconds) are not possible.

9.6 FLOW FEEDBACK (LOAD SENSING)

This is a common method of pump displacement control dating back to the earliest variable volume pumps. The objective is a constant pump output flow regardless of system pressure or pump rotational speed. This is accomplished by passing the entire output flow of the pump through an adjustable orifice, creating a pressure drop, typically 150 psi (Figure 9.7).

The pressures upstream and downstream of the orifice are applied to both ends of a piston. The force on this piston is then applied to the compensator spool. The spool continually meters

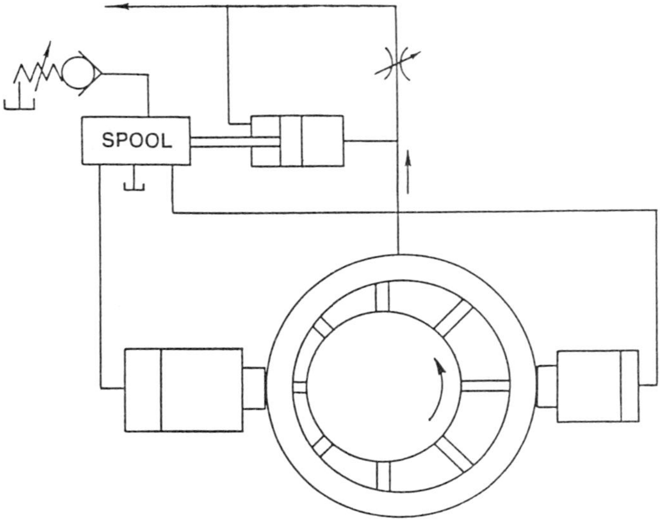

Figure 9.7 Flow feedback control on a variable displacement vane pump.

fluid out of the pump control piston chamber, controlling pressure on the control piston.

As the pump flow increases beyond the desired value, the pressure drop across the orifice increases. The piston then pushes the compensator spool to a compensating position, which meters fluid out of the pump control piston chamber. The pressure drops in this chamber, and the smaller piston pushes the cam ring off-stroke, reducing pump displacement and flow.

As the output flow decreases the orifice pressure drop also decreases and the force of the piston acting upon the compensator spool decreases. With less force on the compensator spool, the spool begins to move back to its original position.

The reverse occurs if the flow drops below the desired value. The pressure drop across the orifice decreases, with a corresponding loss of force on the compensator spool. The spool then meters less fluid out of the control piston chamber and the pressure rises in the control piston chamber. The increased force of the large

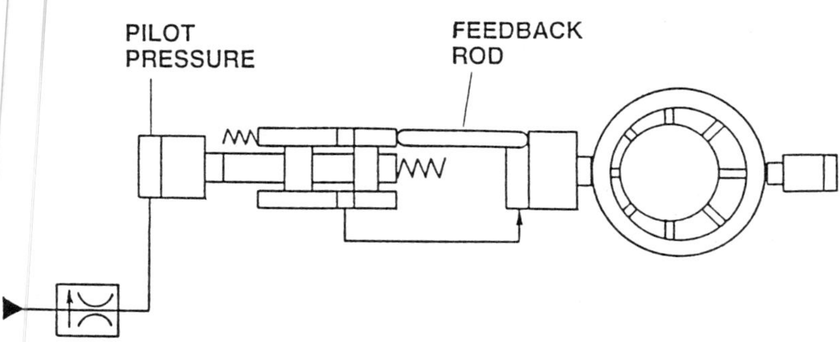

Figure 9.8 Pump displacement control with a follower servo.

pump control piston pushes the cam ring on-stroke, with more displacement and flow.

In summary, the output flow of the pump is controlled by maintaining a pressure drop across an external orifice. The pressure drop feedback (which informs the pump control of the output flow level) is done by a piston acting upon the compensator spool. This approach is accomplished electrohydraulically by using an electrohydraulic flow control valve as the orifice.

9.7 FOLLOWER SERVO METHOD

Another method of electrohydraulic displacement control utilizes electrohydraulic pressure control and mechanical feedback from the cam ring or swash plate. A follower servo valve is a means of providing the position feedback from the pump component that moves with changes of displacement (Figure 9.8 and see Figure 9.11).

The follower servo assembly consists of a spool valve, an outer sleeve, and a feedback rod that maintains a constant distance between the cam ring and the sleeve. Pilot pressure causes the spool to move a distance determined by the force of a precision spring. As the spool moves in relation to the sleeve, ports are unlapped between the spool and the sleeve. Hydraulic fluid then flows through these ports into the control piston chamber. The cam ring is then pushed off-stroke.

As the sleeve moves the same distance as the spool did, the ports between the spool and the sleeve are again lapped, shutting off fluid flow to the control piston. The servo is now in its null position. If the sleeve overshoots the position of the spool, ports are unlapped that vent the control piston.

An electrohydraulic version of a pump like this operates in a similar manner. An electrohydraulic valve controls the pilot pressure by metering fluid out of the pilot pressure chamber. The chamber is supplied by a pressure-compensated flow control valve. In this example the meter out valve is a poppet type, with a solenoid armature exerting force on the poppet.

9.8 ELECTRICAL POSITION FEEDBACK METHODS

An electrohydraulic displacement control method used in some axial piston pumps employs an electrical potentiometer attached to the swash plate to provide electrical feedback of swash plate angle (Figure 9.9). The position signal is sent to the electronic amplifier, which powers an electrohydraulic proportional valve or servovalve.

As a command signal is received by the amplifier, the current is increased to the electrohydraulic valve supplying the control piston. The flow of fluid into the piston chamber is increased, moving the control piston and increasing the swash plate angle and, therefore, pump displacement.

As the swash plate angle increases, the potentiometer rotates and the voltage signal from the potentiometer increases. This signal is fed back to the electronic amplifier, where it is compared with the command signal entering the amplifier. The difference between the command signal and the feedback signal is referred to as an error signal, and the amplifier increases or decreases the current to the electrohydraulic valve according to this signal.

If the swash plate angle is less than the commanded value, the voltage from the potentiometer is less than the command voltage

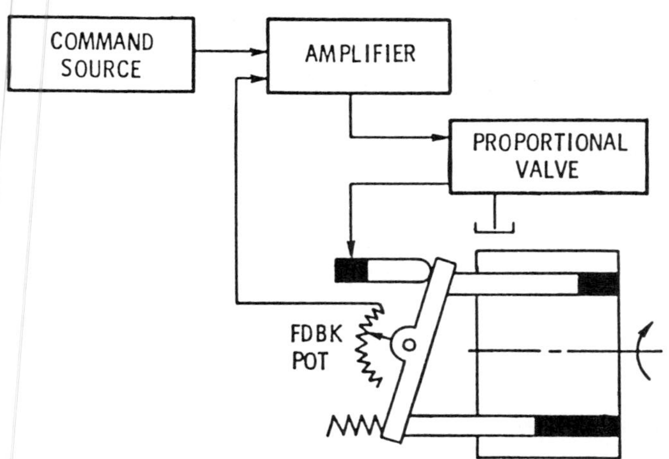

Figure 9.9 Electrohydraulic displacement control with electrical feedback.

and the amplifier sends an increased current to the electrohydraulic valve, causing the control piston to rotate the swash plate further. As the swash plate rotates beyond the commanded position, the potentiometer voltage is greater than the command voltage, the error signal changes in sign (+ to −), and the amplifier sends less current to the valve, reducing the angle of the swash plate.

Linear variable displacement transformers are also used in EH pumps to provide position feedback. The principle of operation is the same as for potentiometer feedback, but the electronic signal conditioning is more complex (see chapter 12).

9.9 DISCRETE PRESSURE AND FLOW CONTROL

The discussion throughout the chapter has regarded proportional pressure and flow control. The pump output pressure or flow was directly proportional to the command signal to the amplifier. There is another means of sequencing a pump along a number of pressure or flow set points—discrete or solenoid-select control.

With this method a number of manually adjustable pressure control modules are connected in parallel. The modules are activated by two-way valves (Figures 9.10 and 9.11). The solenoids of the two-way valves are switched on or off by external switches or discrete outputs from a programmable controller or computer. Note that the computer has control of the sequencing of the setpoint, but it does not have control of the level of the setpoint. The pressure or flow level must be manually adjusted at the control knobs of the pressure modules. This method has the advantage of simplicity; no amplifier is required.

9.10 INTERFACING CONSIDERATIONS

At the present state of the art, interfacing a microcomputer to an electrohydraulic pump is usually done with a D/A (digital-to-analog) conversion at the computer. An analog command signal is usually required by the electrohydraulic amplifiers, which are constructed with analog electronic components. An exception to this are the pumps that contain preset hydraulic modules activated by discrete output from the controller (Figure 9.10).

Microcomputer control can enable the designer to use pump control in designs and aid in exploiting the advantages of control at the pump. The programmability of a microcomputer helps in sequencing actuators to permit control at the pump and in matching pressure and flow set points of the pump to load requirements.

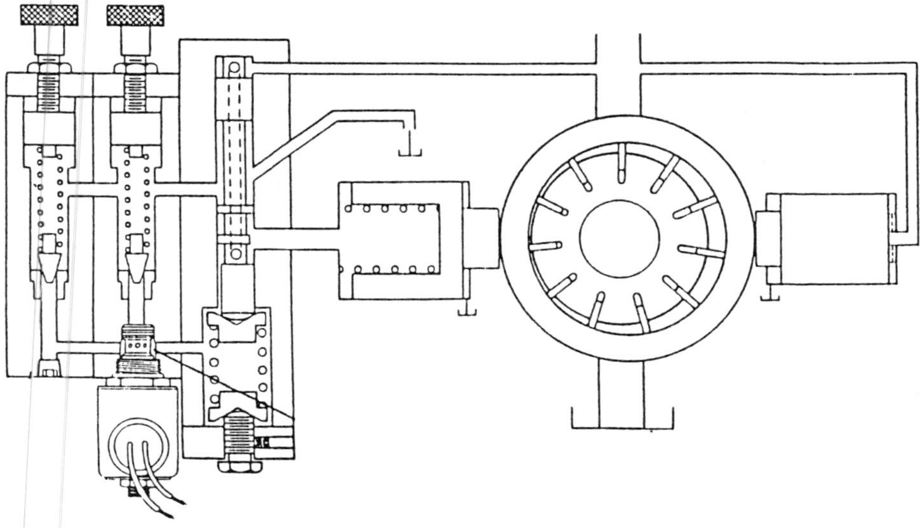

Figure 9.10 Discrete control of two pressures.

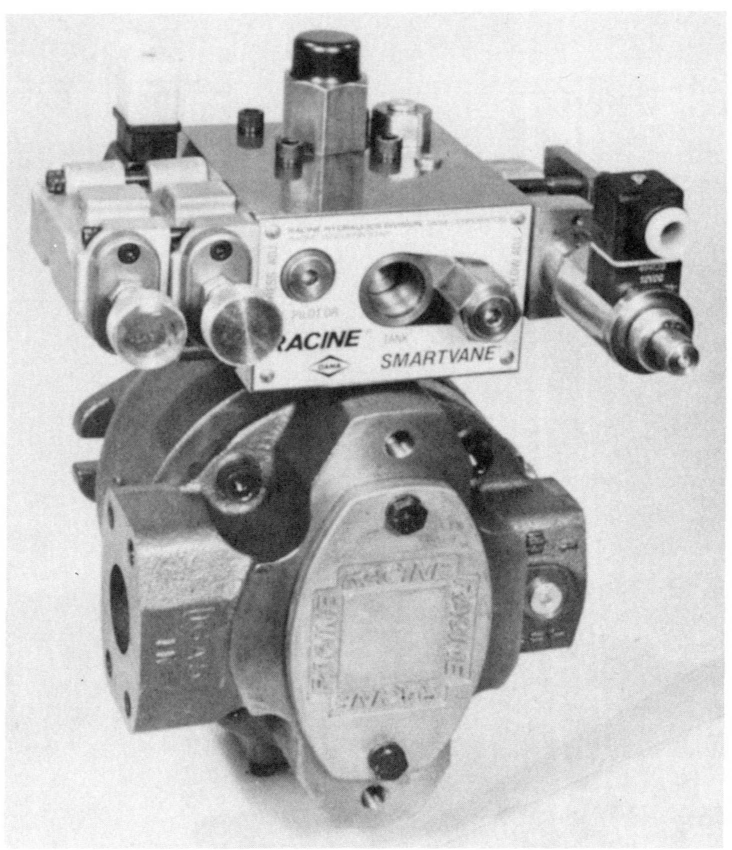

Figure 9.11 Electrohydraulic vane pump with follower servo proportional displacement control and discrete manual adjustable two-pressure control. (SMARTVANE pump courtesy of Racine Fluid Power, Bosch Group, Racine, WI.)

10

Electrohydraulic Actuators (Smart Cylinders)

10.1 INTRODUCTION

"Smart cylinders" is a phrase describing hydraulic cylinders that have some type of transducer that monitors piston position. The position signal can be used a variety of ways by electrical, analog electronic, and microcomputer devices for position control.

10.2 POSITION SENSORS (LIMIT SWITCHES)

The most simple position measurement device, which existed long before smart cylinders, was the electromechanical limit switch (Figure 10.1). It provided a signal of the piston presence at a particular position. The position signal was used for many purposes, such as

Reversing the stroke of a cylinder at the end of its stroke
Beginning deceleration ramping by signaling a proportional valve amplifier card
Stopping a cylinder at a preset position

Proximity switches and photoelectric switches are replacements for electromechanical limit switches (Figure 10.2). These devices use magnetic or light-sensing techniques to detect the presence of an object (such as a piston or cylinder rod). Proximity

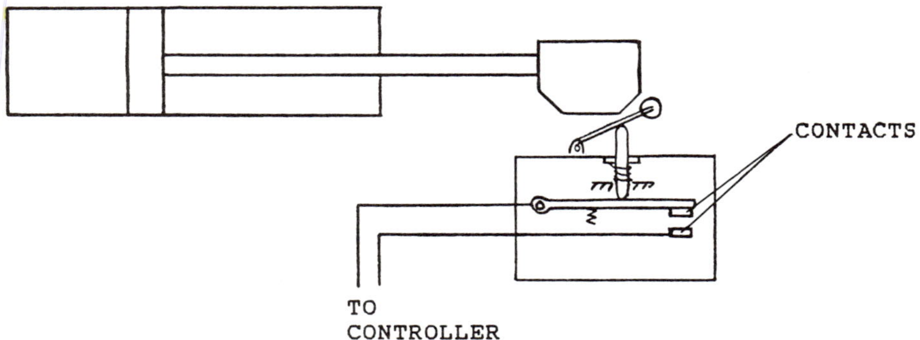

Figure 10.1 Position sensing with an electromechanical limit switch.

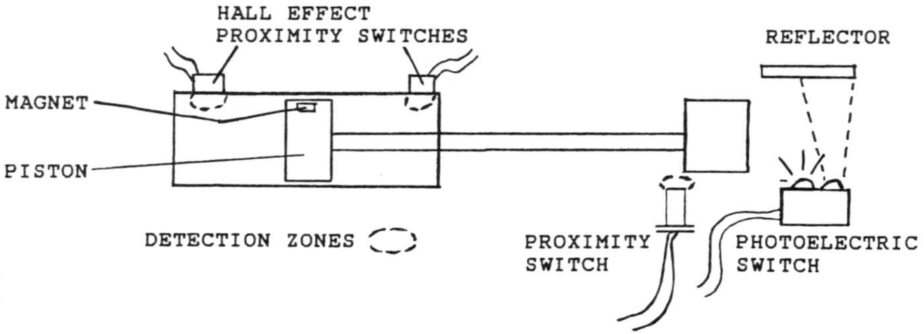

Figure 10.2 Proximity switches and a photoelectric switch.

switches are available that can detect the presence of a steel piston through a cylinder wall (Figure 10.4). These are Hall effect sensors.

The end result is the same with all these sensors. A discrete (on or off) signal is produced when the cylinder piston is at a specific stroke.

10.3 POSITION SENSORS (CONTINUOUS)

Continuous position sensors detect piston position at an infinite number of positions throughout the stroke of the cylinder. The electrical signal varies in proportion to the distance of the piston from its retracted position. A single continuous position-sensing device can replace many limit switches. To do this an electronic amplifier or microcomputer is required to receive this signal and perform whatever action is dependent upon the cylinder position.

10.3.1 Potentiometers

A simple continuous position transducer, which was also in use before "smart cylinders" existed, was the linear potentiometer (Figure 10.3). A variation of this was the reel and rotary potentiometer (Figure 10.4).

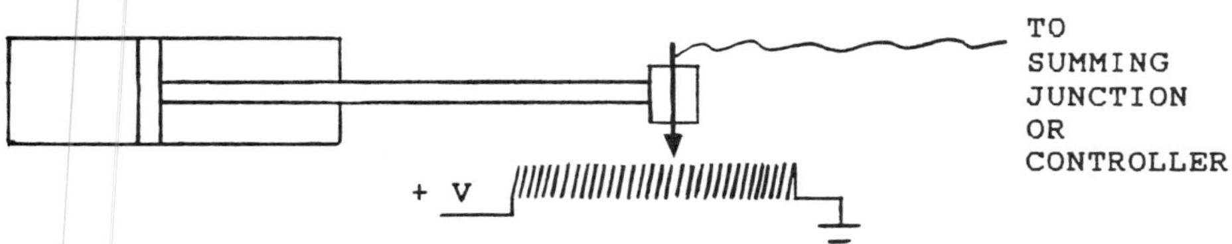

Figure 10.3 Position feedback by a linear potentiometer.

Linear potentiometers have been used in closed-loop position control systems for many years. In these systems a proportional valve or servovalve is used to control the stroke of a cylinder and the position signal enters an analog electronic amplifier to be summed with the commanded position signal.

10.3.2 Advances in Position Transducers

The microcomputer, along with the development of highly sophisticated position transducers, considerably advanced the state of the art of cylinder positioning. The positioning accuracy of cylinders that utilize these advanced techniques are referred to as "smart cylinders."

One of these new methods utilizes a cylindrical rod with special machining (Figure 10.5). The cylindrical rod has a series of grooves cut into its surface. The rod is then plated with a nonmagnetic metal, such as chrome. The rod is then remachined to its correct diameter. Magnetic sensors, such as Hall effect sensors, detect the passing of the grooves and count them. This sensor is incremental and does not provide an absolute reading of cylinder

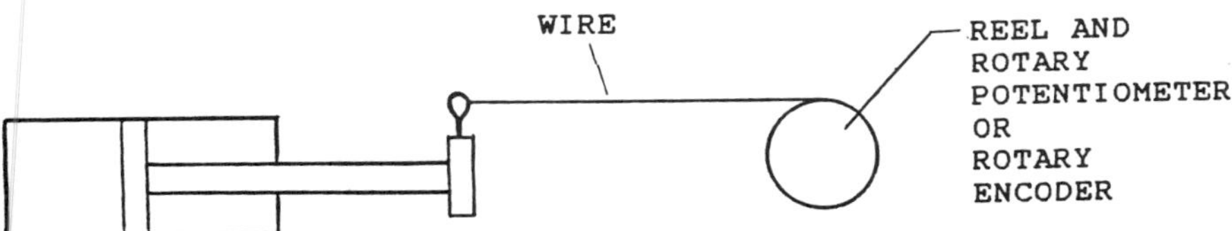

Figure 10.4 Position feedback by conversion to rotary motion.

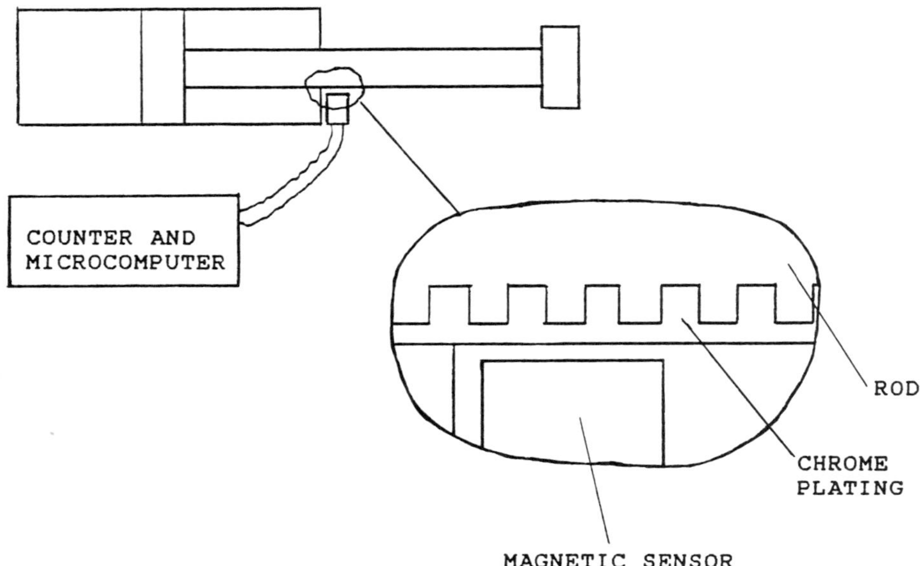

Figure 10.5 Magnetic, incremental sensing of cylindrical rod position.

position. Absolute positioning can be done, with software keeping track of the stroke, from a starting (indexed) position.

Another method uses ultrasonic waves to detect piston position. The ultrasonic waves are emitted at one end of the cylinder and travel through the hydraulic fluid. Upon contacting the piston, the waves bounce off the piston and return to the wave generator. The time for the wave to return is precisely measured, resulting in an accurate indication of position.

A positioning system that has attracted much attention in connection with smart cylinders is the Temposonic system, manufactured by the Temposonics Division of MTS Systems, Inc. (Figure 10.6) [1]. It operates on a magnetostrictive principle.

A wire is mounted in a tube that passes through the center of a cylindrical rod. Electric current pulses are sent through the wire. A magnet is mounted on the piston of the cylinder. The pulse interacts with the magnetic field of the magnet and deflects the wire

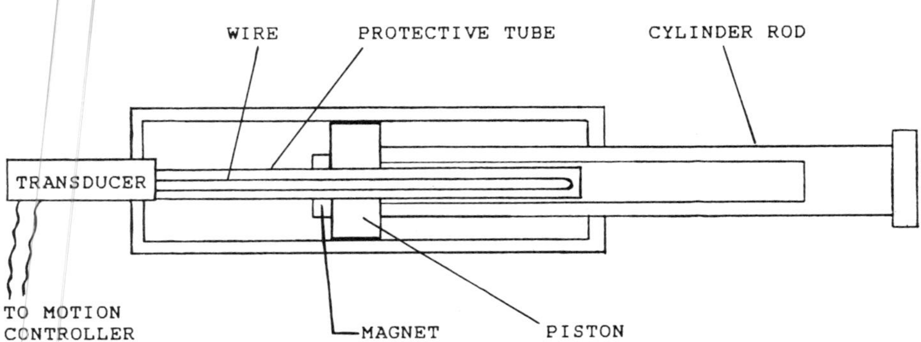

Figure 10.6 Piston position measurement by a Temposonic magnetostrictive transducer.

torsionally (twisting it slightly). The electronics of the transducer measure the time between the sending of a current pulse and receiving the torsional reaction. Piston position is determined to an accuracy of less than .001 inch.

10.4 POSITIONING SYSTEMS

To fully exploit the capabilities of transducers like the Temponsic, hydraulic manufacturers believed that they had to provide a complete positioning subsystem or package. The cylinder with integral transducer, servovalve, servovalve electronic amplifier, and the motion controller had to be sold as a package. With these packages, the hydraulic industry had positioning systems similar to the step motor positioning packages used by the electronics industry. The similarity of the systems is shown in Figures 10.7 and 10.8.

Both are closed-loop positioning systems. Both use microcomputer-based controllers to close the position loop. This type of control system is widely used with step motors. The microcomputer controller in these step motor applications is referred to as a motion controller. The motion controller closes the position loop and calculates the present position of the piston, future position, and acceleration profiles to reach these positions.

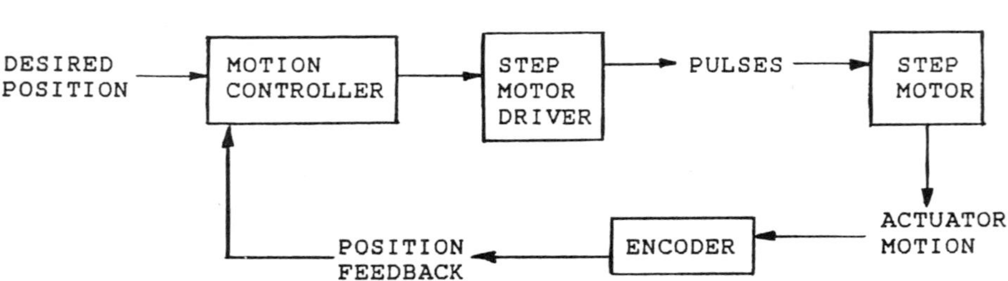

Figure 10.7 Position control system with step motor.

10.5 RECENT ADVANCES

A microcomputer is used by Vickers [2] of Troy, Michigan, to improve the response time of their smart cylinder. A microcomputer is housed within the cylinder and servovalve assembly.

Using dynamic data from the system to be controlled, the microcomputer can increase the speed of response of the cylinder without inducing instability. This system is on the "cutting edge" of microcomputer control of hydraulic systems. It is an example of the microcomputer enhancing the dynamic response of a system.

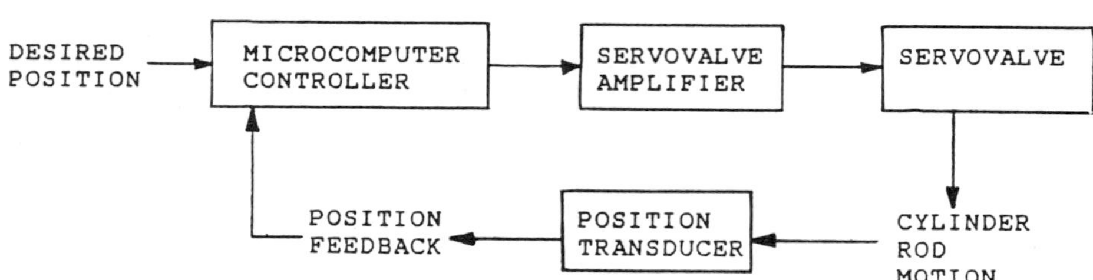

Figure 10.8 Position control system with a microcomputer closing the position loop.

REFERENCES

1. Temposonics, a division of MTS Systems Corporation, 131 Ames Court, Plainview, NY 11803.
2. Vickers, a division of the Trinova Corporation, 1401 Crooks Road, Troy MI 48084.

11

Sensors

11.1 INTRODUCTION

Sensors perform the task of converting a physical phenomenon into electrical signals. A microcomputer can then receive these signals, interpret them, and send command signals to a hydraulic system to change a set point or sequence.

Hydraulic system parameters that might be useful to the controller are cylindrical rod velocity, hydraulic motor rotational speed, pressure, cylinder position and speed, and temperature. In this chapter, the conversion of these phenomena into useful information for the microprocessor is discussed.

11.2 PRESSURE

Pressure in a hydraulic system can be monitored by pressure switches or pressure transducers.

11.2.1 Pressure Switches

A pressure switch consists of a diaphragm exposed to system pressure and a linkage attaching it to a two-position electrical switch (Figure 11.1). As the pressure of the system exceeds a present value, the switch is tripped and the contact closure of the switch can be used for indicator lights or for input into a microcomputer. Pressure switches are also available that are activated by a lower than expected pressure or for indicating that pressure is between two presets.

11.2.2 Pressure Transducers

In a pressure transducer, the pressure value is converted into an electrical voltage proportional to the pressure. The transducer typically consists of a metal diaphragm with a strain gage bridge bonded to it (Figure 11.2). As pressure increases, the diaphragm is deflected, causing a change in the resistance of the strain gage

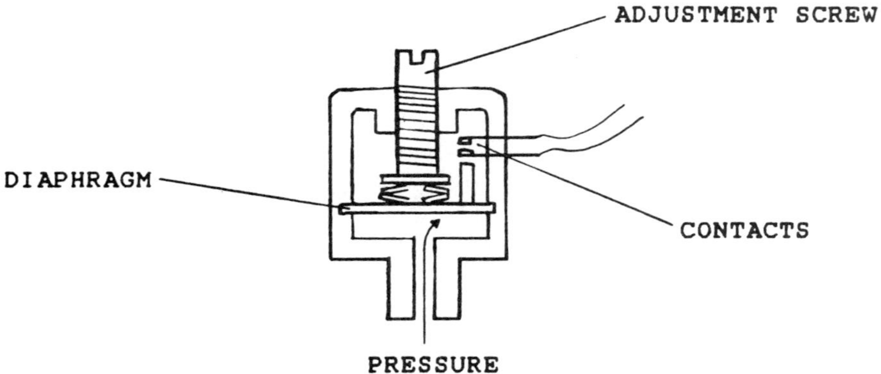

Figure 11.1 Pressure switch.

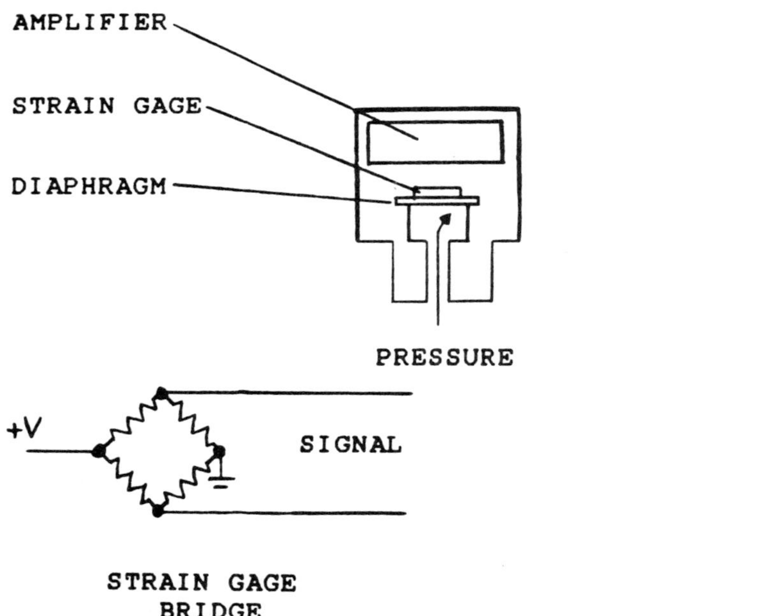

Figure 11.2 Transducer and strain gage bridge.

bridge. Therefore, the voltage drop across the bridge is proportional to the pressure applied to the diaphragm.

The voltage drop is small, less than a volt, and must be amplified for most applications. Transducers are available with amplifiers built into the transducer housings. Some of these are referred to as pressure transmitters. These transducers have a standard output of 4–20 mA or 0–5 V. If a transducer must be used that does not have an amplifier and the output is in millivolts, an amplifier circuit can be constructed (see Figure 5.15).

Once the output is in a standard range (0–5 V, 0–10 V, 4–20 mA) it can be converted to a binary number by an analog-to-digital (A/D) converter. The resulting number can be sent to an input port of the microcomputer.

At the present time there are not completely digital pressure transducers. Commercial transducers may exist with digital output, but these must go through an analog-to-digital conversion. The reason for this lies in the only two ways a microprocessor can receive information: either a binary number, or a time interval between events. At the present time there is no way to convert a pressure into two events, with the time interval proportional to pressure.

11.3 TEMPERATURE

Temperature transducers are usually one of three types—thermocouples, resistance temperature detectors (RTDs), and thermistors.

Thermocouples consist of a junction of two wires of dissimilar materials. A voltage output is produced that is proportional (but not linearly) to temperature. Thermocouples are the most widely used temperature transducer.

Resistance temperature detectors are wire-wound resistors. The resistance increases with temperature. Compared with thermocouples RTDs are more expensive and more linear, have a longer life, and do not require the "cold junction" reference point that thermocouples need.

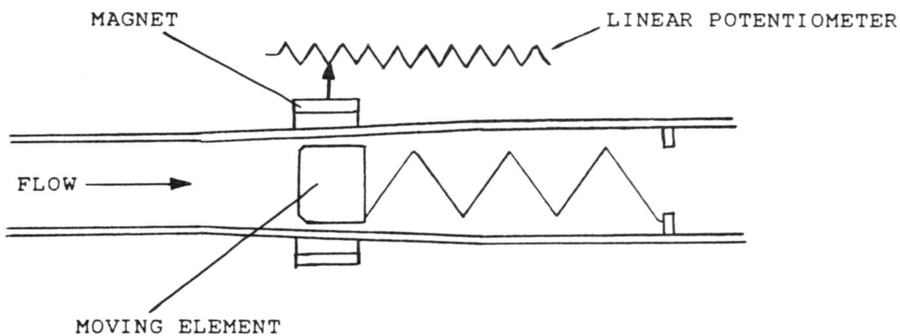

Figure 11.3 Variable area flow meter with linear potentiometer.

Thermistors are made from sintered metal oxides. The resistance of these devices changes with temperature. The resistance change is exponential with temperature, requiring more complex signal conditioning. The advantage over the other types is the repeatability from unit to unit.

All three types of temperature transducers produce an analog signal. The voltage produced by these transducers must be converted by an analog-to-digital converter for use by a microcomputer.

11.4 FLOW

The flow rate of hydraulic fluid is usually measured by one of two methods, variable area or turbine flow meters.

Variable area flow meters (Figure 11.3) have an element that moves to a position in a tube proportional to the flow rate through the tube. A magnet mounted outside the tube follows the element along the length of the tube. A linear potentiometer can be used to convert the position to a voltage.

Turbine flow meters (Figure 11.4) use a rotor consisting of five to eight blades that resembles a fan or ship's propeller. The rotational speed of the rotor is proportional to the rate of fluid flow. A magnetic pickup is used with these devices to send a pulse with the passing of each blade. These are very accurate and are suitable for

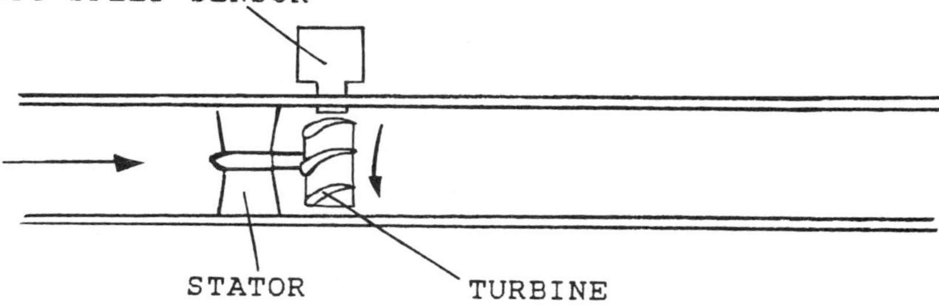

Figure 11.4 Turbine flow meter.

laboratory work. The pulses can then be conditioned and counted by a digital counter or transformed into an analog voltage by a frequency-to-voltage (F/V) converter.

However, flow meters have several inherent disadvantages for use in control systems. Turbine flow meters, which are the most accurate, are viscosity sensitive. The readings must be converted for the temperature of the hydraulic fluid.

Another disadvantage is that a flow meter does not sense slip or leakage in an actuator. The speed of an actuator, like a hydraulic motor, decreases owing to slip. (Slip is defined as internal leakage around a seal of a piston from the high-pressure side to the low-pressure side.)

This brings up the important point that actuator speed is more important to the system designer than fluid flow rate. The system designer is more accurate in the control of a system if he or she senses actuator speed directly with a speed transducer than by calculating the speed by using fluid flow rate. This is the reason that speed measurement is discussed here instead of flow rate measurements.

11.5 SPEED TRANSDUCERS

The rotational speed of hydraulic motors can be measured by several methods—tachometer generators, magnetic sensors, and encoders.

11.5.1 Tachometer Generators

These are miniature dc generators that output a voltage proportional to the rotational speed. This has been a popular method for years but is now being replaced in many applications by encoders and magnetic sensors. The encoders and magnetic sensors offer greater accuracy than tachometer generators at a similar price.

Tachometer generators offer more accurate response to quickly changing speeds than pulse sensors. This is because of the time that it takes for the counting circuits of the F/V converters or encoders to count a pulse train and calculate an average speed. The change in output voltage from a tachometer generator is instantaneous with a change in rotational speed.

This is necessary in closed-loop speed control systems. In such systems encoders with high pulses or revolution must be used, which can increase the cost of the system.

11.5.2 Magnetic Pickups

Magnetic pickups contain a permanent magnet surrounded by a conductor. The magnet is placed in close proximity to a gear mounted on a shaft (Figure 11.5). Each time a gear tooth enters the magnetic field of the sensor, a voltage is generated in the conductor. The series of voltage pulses can then be conditioned to produce digital pulses, which can then be counted by digital counters. If an analog voltage reading of speed is desired, a frequency-to-voltage converter can be used with the magnetic pickup.

11.5.3 Incremental Encoders

Speed can also be monitored by optical encoders. These devices use a disk that rotates with the shaft to be monitored (Figure 11.6). A light source supplies light to one side of the slotted disk. A photovoltaic sensor reacts to the pulses of light as the shaft turns and produces voltage pulses.

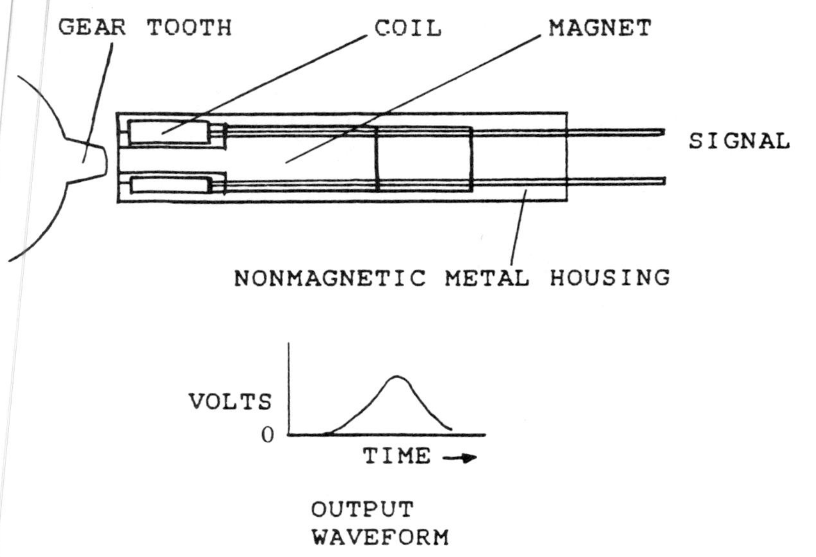

Figure 11.5 Magnetic sensing of shaft speed.

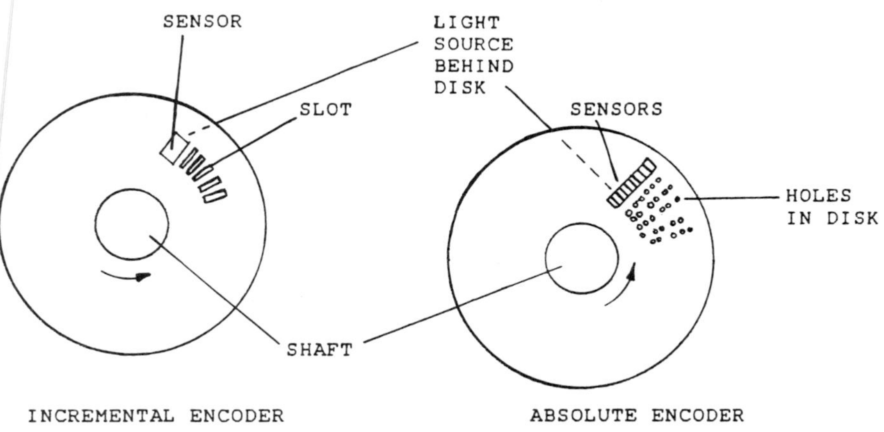

Figure 11.6 The disks used in two types of optical encoders.

The pulses are then counted prior to sending this information to the input port of a microcomputer. This can be done by a counting circuit (Figures 6.9 and 6.10) or by counters commercially available for this purpose. The resulting binary number is sent to the microcomputer's input port.

11.6 POSITION SENSORS: ABSOLUTE ENCODERS

These are primarily position sensors. The difference between these and incremental encoders is the slots in the disk. Several rows of slots are used. The slots are arranged from row to row in such a way that the resulting openings form a code. Each shaft position has a unique combination of slots open to the light source. A number of photovoltaic sensors detect the openings in the various rows of slots. Upon start-up, an absolute encoder sends a signal of the present position to the computer. Absolute encoders are more expensive than incremental encoders and require more expensive signal conditioning circuitry than incremental encoders. Speed measurement can also be done with an absolute encoder, but a calculation algorithm is required in software to calculate actuator speed.

11.7 SIGNAL CONDITIONING

The voltage signal from a transducer is often too low to use with A/D converters. It must be amplified to the proper range and biased to match the low point with the A/D converter requirements. This is discussed in Chapter 5, Figures 5.14 and 5.15.

Fundamentally, these circuits work but are not considered state of the art. For those who desire to purchase components specially designed for signal conditioning applications, the 5B series of devices from Analog Devices Inc. (see Reference 3, Chapter 6) should be considered. These modules provide additional benefits, such as transformer isolation and greater accuracy.

12

Microcomputer Control of Hydraulic Systems

12.1 INTRODUCTION

Early hydraulic systems were controlled by manually operated valves and human operators. Simple automated sequences could be accomplished through hydraulic logic. Valves could be arranged so that a pressure rise in one valve would activate another.

Electrohydraulics brought electrically operated valves and remote control. Valves could be switched from one position to another with two-position electrical switches. Remote set point adjustment of pressure and flow control valves was also made possible. This could be done with potentiometers on a control panel.

Microprocessor control is another revolution. The microprocessor has improved the accuracy and efficiency of hydraulic systems and has added "intelligence." The hydraulic system can now monitor its operation, adjust itself for changing conditions, and communicate the results to management's computers!

In this chapter, the benefits of microcomputer control of hydraulic systems are discussed.

12.2 PROGRAMMABILITY

Programmability is the capability of storing a sequence of events that are then executed automatically by a control system. The program can be a simple sequence that repeats time after time, like a tape recording or a player piano. The program can also be complex to the point that it has decision-making capability. It can accept such inputs as speeds or pressures and can change the execution of a sequence.

12.2.1 Hydraulic Programmability

Hydraulic systems had programmability before microcomputers and even before electronics were applied. Hydraulic logic was used. Figure 12.1 is an example of a hydraulic system that produces a three-speed sequence of a cylinder. The speed sequence is

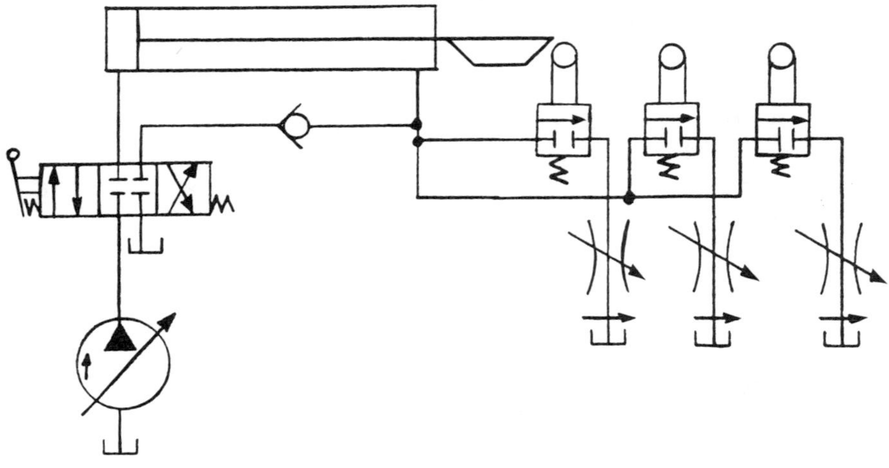

Figure 12.1 Speed sequence by hydraulic control.

automated as the cylinder extends. The speeds are adjusted by three flow control valves.

12.2.2 Electrical Programmability

Electrically operated hydraulic systems also had programmability (Figure 12.2). In this system the cam-operated valves of the first example are replaced by electromechanical switches. The flow control valves are replaced by an electrohydraulic proportional valve. The switches activate control potentiometers connected to the amplifier of the proportional valve.

The first two examples repeated a simple sequence, but consider the changes necessary to change the sequences of these systems. If the sequence is changed in Figure 12.2, the cam and the position of the cam-operated valves may change. This requires replumbing the hydraulic circuit. Additional flow control valves must be added to provide additional speeds.

In Figure 12.2, the location of the switches must change to change the timing of the sequence. However, additional speeds

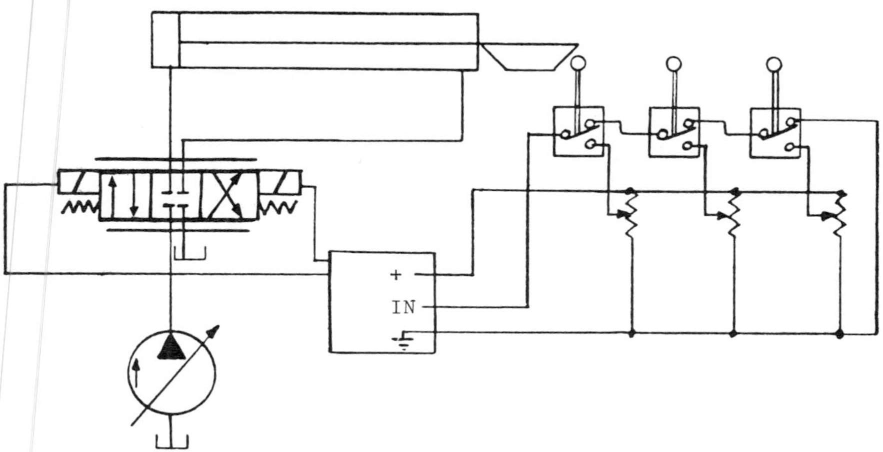

Figure 12.2 Speed sequence by electrohydraulic control.

require only another limit switch and potentiometer. (This shows the value of an electrohydraulic flow control.)

12.2.3 Microcomputer Programmability

A system with a microcomputer-controlled three-speed sequence is shown in Figure 12.3. Two of the switches have been eliminated. The speed sequence is controlled by timers in the program of the PLC. This method has the following advantages over the hydraulic and electrohydraulic examples:

Circuit simplification. No longer is a flow control valve or limit switch required for each flow rate in the cycle. Adding another speed to the sequence can be done with a software change only. A timer can be programmed in the computer that adds another speed to the sequence.

Ease of changing the sequence. Because it is not necessary to move any of the components to change the sequence, this can be done quickly and conveniently. This shortens the set up time of a machine (making smaller lot size production practical). Not only is the time saved in moving the components, but

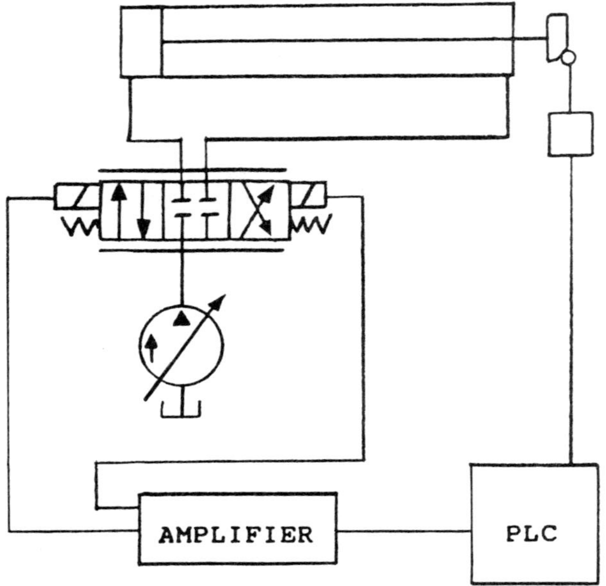

Figure 12.3. Speed sequence by microcomputer program.

the time spent adjusting the components for precise operation is also eliminated.

Maintenance is reduced with a microcomputer-controlled system. The solid-state logic does not wear and fail as often as electromechanical components. Therefore, troubleshooting time to locate worn components is also reduced.

12.2.4 Special Effects and Functions

Special control effects are made possible by programmable set point control. Acceleration of the cylinder in Figure 12.3 can be accomplished by ramping of the flow set point. In ramping, the flow set point is continually changed in small steps from one flow rate to another (Figure 12.4).

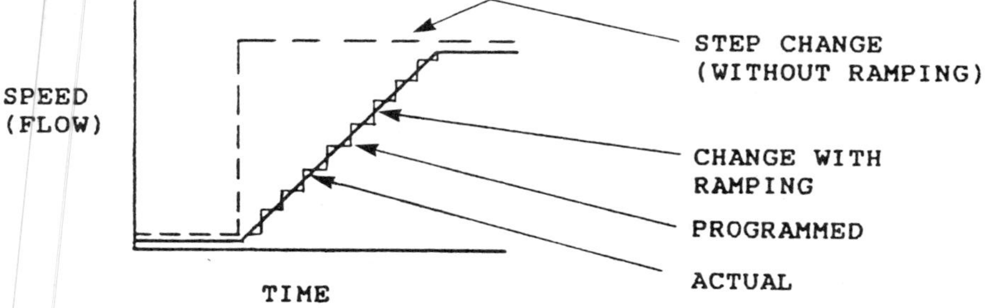

Figure 12.4 Acceleration control.

Ramping can be accomplished in electrohydraulic circuits without microcomputer control, but a programmed ramp has several advantages.

No hardware is required. Ramping circuits, in the amplifier or external, are not required.

An acceleration curve of any shape desired can be programmed. For optimum cycle times with maximum acceleration and minimum shock loading to a machine, constant acceleration is required. Some electrohydraulic components do not have constant acceleration capability in the electronics.

A large number of different acceleration profiles can be programmed into a PLC. Accomplishing multiple ramp speeds with electrohydraulic components and discrete switches can result in a complicated command circuit with reliability and maintenance problems.

The capability of programming calculation of a set point is an important advantage that microcomputers bring to the control of hydraulic systems. An example of this is in a plastic injection molding machine (Figure 12.5). In this process, hydraulic pressure and flow must be precisely controlled. Because of moisture variations in the bulk plastic material, it may be necessary to change the pressure and flow values from one batch to the next. Process controllers for injection molding have an offset command in the software that allows the operator to change all the set points by a few percent with one command. The controller calculates the new preset values and modifies the analog signal to the amplifier.

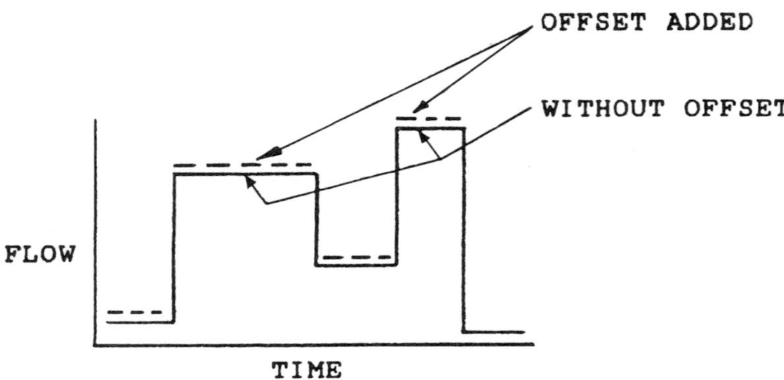

Figure 12.5 Example of a calculated set point.

12.2.5 Energy Conservation Through Programmability

The energy consumption of a fixed displacement, constant flow hydraulic system can be reduced by an electrohydraulic relief valve controlled by a microcomputer (Figure 12.6).

In this system, the relief valve controls the pressure. The output flow of the pump is constant at the maximum available from the pump, 10 gallons/minute. The excess flow not required for the load is directed to tank at system pressure. This is wasteful of energy owing to the bypassed flow and the pressure drop across the relief valve. If the relief valve can be programmed to a pressure near the load pressure, a savings in energy can be realized. A mismatch of relief valve setting to the load can occur in systems that handle both large and small loads. Without the option of programmability, the relief must be set at a pressure sufficient to handle the largest loads.

An improvement over the constant flow, fixed displacement systems are the constant pressure or demand systems (Figure 12.7). These systems utilize a variable volume pump. Pressure is

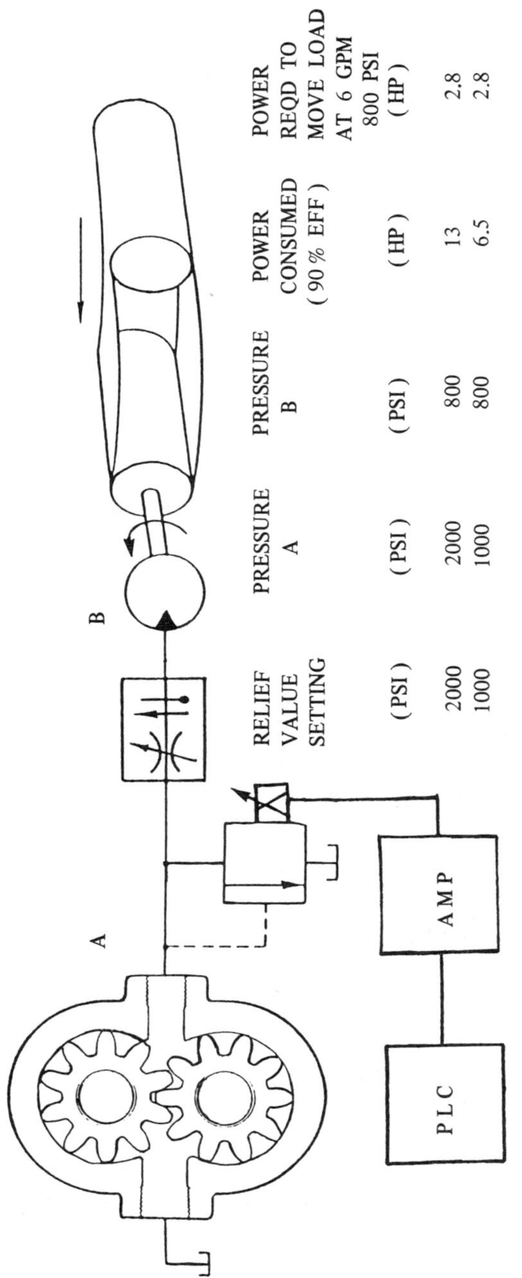

RELIEF VALUE SETTING (PSI)	PRESSURE A (PSI)	PRESSURE B (PSI)	POWER CONSUMED (90% EFF) (HP)	POWER REQD TO MOVE LOAD AT 6 GPM 800 PSI (HP)
2000	2000	800	13	2.8
1000	1000	800	6.5	2.8

Figure 12.6 Energy efficiency by relief valve programming.

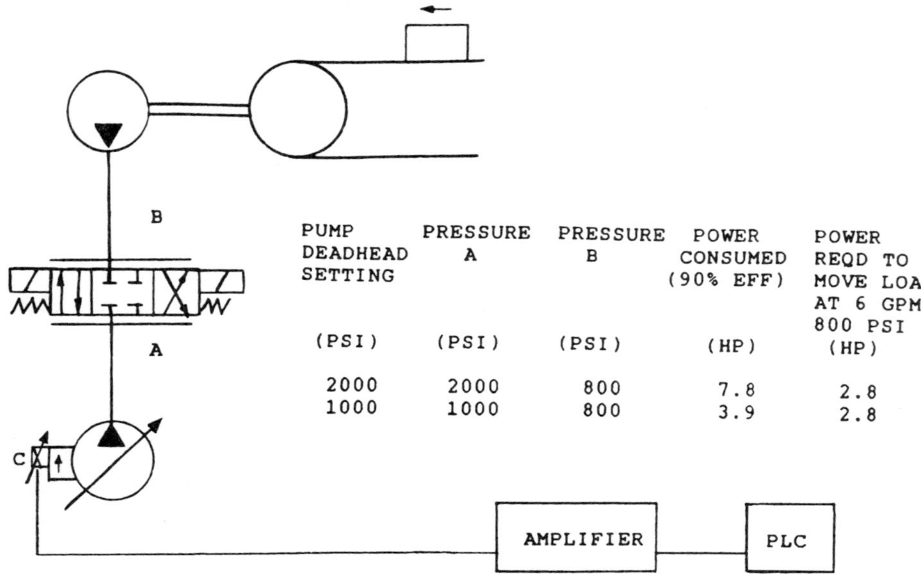

PUMP DEADHEAD SETTING (PSI)	PRESSURE A (PSI)	PRESSURE B (PSI)	POWER CONSUMED (90% EFF) (HP)	POWER REQD TO MOVE LOAD AT 6 GPM 800 PSI (HP)
2000	2000	800	7.8	2.8
1000	1000	800	3.9	2.8

Figure 12.7 Programming pump deadhead setting to match the load for energy efficiency.

set at the pump, and the pump automatically reduces or increases pump displacement to maintain constant pressure in the system. The output flow rate from the pump is only enough to maintain system pressure so that there is no excess flow that must be bypassed to a tank. The reduction in pump flow results in energy savings and lowered heat exchanger requirements compared with constant flow systems. The energy consumption of these systems can also be improved through programmable control.

One method would be to program the computer to control the pressure setting of the pump. If the pressure setting of the pump can be programmed as slightly above the amount required by the load, the pressure drop across the proportional flow control valve is reduced.

Another method that improves the energy consumption of this system is to program the displacement of an electrohydraulic pump (Figure 12.8). In this system the flow control valve is elimi-

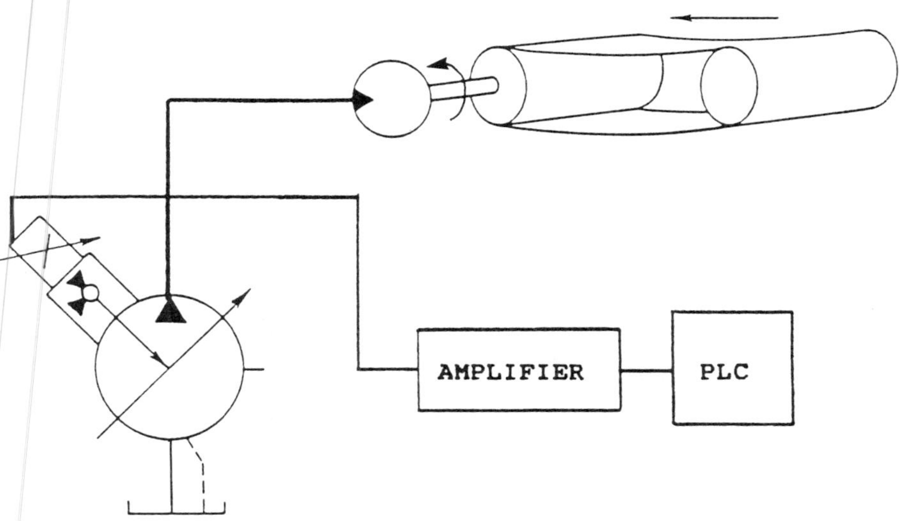

Figure 12.8 Programming the displacement of a pump.

nated. The flow required to produce the desired actuator speed is programmed. The pump produces the commanded flow, and the pressure rises to whatever the load resistance was. This is perhaps a better example of an electrohydraulic pump than programmability. However, if different speeds are desired for different products, programmability can be used to optimize this system for energy consumption.

12.3 CLOSED-LOOP SYSTEMS

Closed-loop control has been used in process industries and hydraulic systems for many years. It is done by placing a transducer in the system to monitor a parameter in the system, such as speed or pressure (Figure 12.9). The signal from the transducer is sent to a circuit referred to as a summing junction. In the summing junction, the feedback signal is continuously compared with the command signal, and a difference is amplified and added or subtracted from the command signal to the amplifier. This has the result of

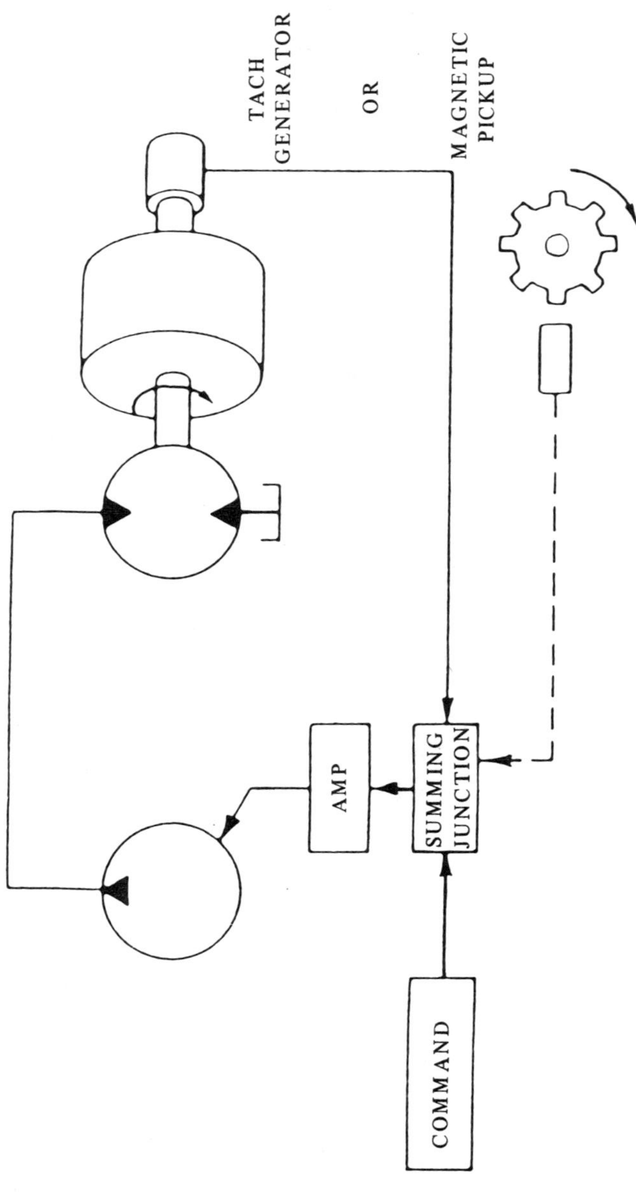

Figure 12.9 Closed-loop flow control by pump displacement control.

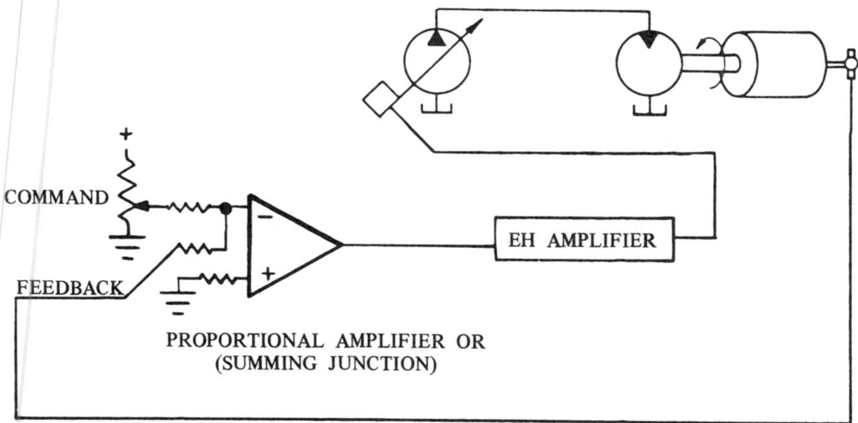

Figure 12.10 Closed-loop control with a proportional amplifier.

providing the proper amount of pump output flow to bring the motor back to the desired speed.

An example of this is shown in Figure 12.9. If the motor is at a speed of 50%, the load on the motor increases, and leakage in the pump and motor increases, the motor speed slows to perhaps 46%. With a closed-loop system the difference (4%) between the motor speed and the commanded speed is sensed at the summing junction, and an increased command signal (4% or more) is sent to the amplifier that controls the displacement of the pump. The increased flow (54% or greater) from the pump brings the motor back to the desired speed of 50%.

12.3.1 Proportional Controllers

The most simple type of closed-loop controller is referred to as a "proportional" controller. It consists of an operational amplifier circuit (summing junction), as shown in Figure 12.10.

A proportional controller with a gain of 1 was discussed in the example of Figure 12.9. The controller increased its command to the pump by an amount equivalent to the error. The command to the pump was changed by 4% for an error of 4%. The gain of a proportional controller can be changed. A gain of more than 1 will result in increased sensitivity and less than 1, in decreased sensitivity.

A disadvantage in a system controlled by a proportional controller is steady-state accuracy. Because the correction of the output of the system is dependent only on the error between the feedback and the command, there is always an error or difference between the command and the output (Figure 12.11).

Another disadvantage of a proportional controller is system instability. The controller is sensitive to the dynamic response of the system. The response of the pump, the hydraulic capacitance of the system, the response time of the pump, and the electronics can combine to produce an unstable or oscillating system. The instability can be eliminated by electronic damping, but this slows the response of the system.

12.3.2 PID Loop Controllers

PID (proportional, integral, and derivative) loop controllers solved most of the problems of proportional controllers. The proportional part of this has the same function as in the proportional controller.

The integral function of the controller continually integrates the difference between the desired speed and the actual speed and adds or subtracts this from the set point. The integral function prevents the steady-state error of a proportional controller.

Another function of a PI controller is reset action. The controller calculates a set point and outputs the set point value without change for a period of time, for example, .1–10 s. By holding a single value for a period of time, the controller is not as sensitive to the dynamics of the system (Figure 12.12). Instability is easier to control than with a proportional controller.

The derivative function adds another enhancement to the proportional controller. It responds to the rate of change of the controlled variable. If an abrupt, rapid change occurs in the speed, the controller begins a correction immediately. It does not wait for the reset interval to time out. The effect of the derivative function is to

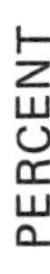

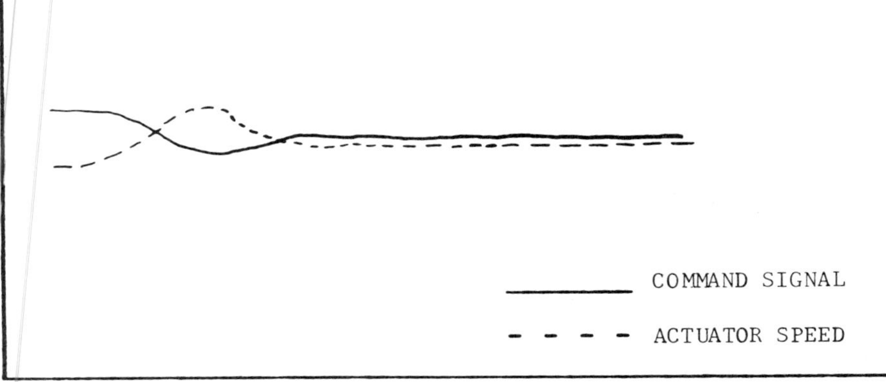

Figure 12.11 Steady-state error in a proportional control system. (Percent axis refers to actuator speed and signal from summing junction to electrohydraulic amplifiers.)

reduce the error in a system that is correcting for a large disturbance.

PID controllers have been used for 30 years. Until a few years ago these were analog devices (built with analog circuitry). PID controllers with analog circuitry proved to be difficult to adjust and "tune" in the field. This was due to the characteristics of the analog circuitry. The analog circuitry also had drift problems that degraded control accuracy.

Digital PID controllers substituted digital circuitry, a microprocessor, and software for the PID calculation. The result was that tuning was easier. The software algorithms provided this. Another advantage was that drift was eliminated. Digital numbers do not change in value with temperature.

An important advantage and breakthrough that the microprocessor brought to the PID controller was "adaptive" control. This is also referred to as "self-tuning." The controller, as it operates, monitors how well it is "guessing" the correct set point to maintain the desired speed. The controller readjusts its P, I, and D settings to

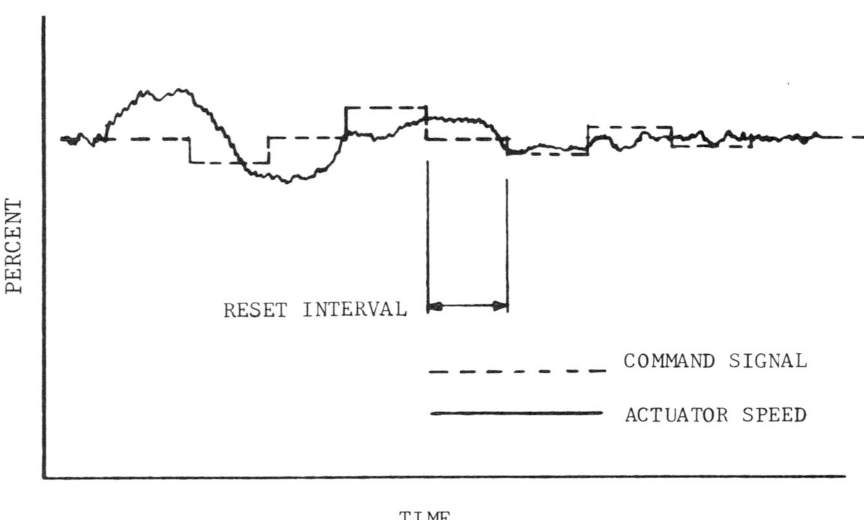

Figure 12.12 Reset action of a PID loop controller. (Percent axis refers to actuator speed and signal from PID loop controller to electrohydraulic amplifier.)

improve its control accuracy. Just as a human operator does after hours of practice, the controller readjusts its own responses to improve its performance.

12.4 DECISION TO USE MICROCOMPUTER CONTROL

A hydraulic system designer must eventually answer the question: Should I use a microcomputer to control my hydraulic system? The answer to this question is becoming "yes" in more and more systems because of the fallings costs of microcomputers and the growing number of people familiar with these controllers. The following is a discussion of the trade-offs to be analyzed before deciding on a control method for a hydraulic system. Hydraulic, electrohydraulic, and microcomputer control methods are discussed.

Hydraulic control systems are considered those that use hydraulic valves for the control logic (Figure 12.1). Electrohydraulic systems are those that use electrical or electronic controls for both discrete and proportional valves and pumps. The control logic is in electrical hardware, such as limit switches and relays (Figure 12.2). Microcomputer control systems use a microcomputer, such as a PLC, board-level, or industrial microcomputer, to control the sequence of a hydraulic system (Figure 12.3).

12.4.1 Initial Cost

An electrohydraulic (EH) system can have an advantage over a hydraulic control system if numerous flow or pressure control valves can be replaced by a single electrohydraulic proportional valve. If an electrohydraulic control system has more than four electromechanical relays, a PLC can be a cost-effective alternative.

In a PLC system, the software costs usually exceed the cost of the controller. Translating the cycle into a software program can be a time-consuming task if the programmer is inexperienced. The important point is that once a duty cycle is defined, consultants are available for programming a PLC at a reasonable rate. (The design of the duty cycle must be done regardless of the control system used.) Wiring a PLC into a system can be easier and more quickly done than installing electromechanical relays. The end result is usually neater in appearance and easier to service with a PLC.

If a sequence of presets is required, the cost trade-off between a PLC with analog output and the combination of a PLC with discrete output and external control relays must be weighed. PLCs with discrete I/O are very inexpensive. Wiring of the analog output PLC system is simpler, as Figure 12.13 shows. However, BCD thumbwheel switches and input modules are more costly than potentiometers.

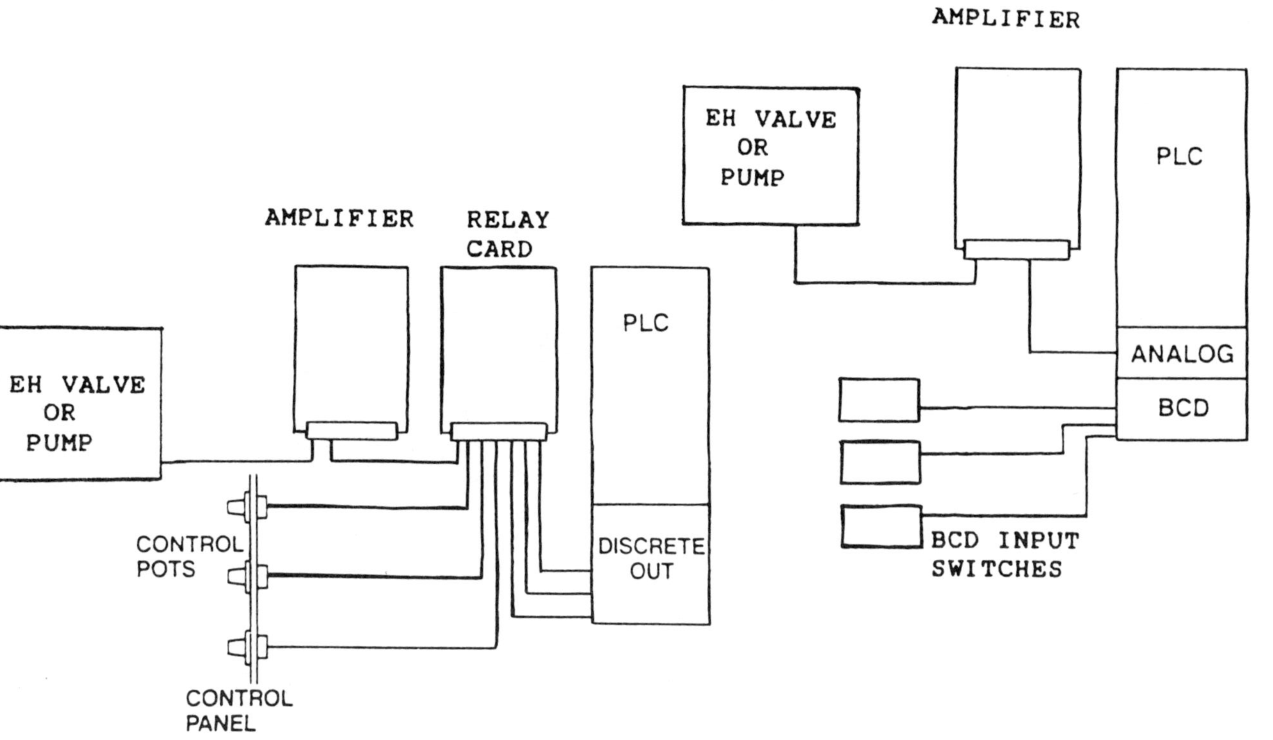

Figure 12.13 Comparison of command circuit PLC with discrete or analog output.

12.4.2 Frequency of Adjustment

How often will the duty cycle be adjusted? This refers to adjustment for timing of the sequence and set point adjustments. If the duty cycle will not change, hydraulic control has an advantage in simplicity and reliability over the other methods.

If the set point adjustments must be changed often, all three methods can be used. The selection depends upon other factors, such as precision of the adjustment and convenience. If the set point adjustments will be made often and are tricky or time consuming, the programmability of an analog PLC systems saves time and effort.

If the sequence of the duty cycle will be changed often, for example, when running different parts through the same machine tool, the programmability of a PLC is a great advantage. Changing the sequence of a hydraulically controlled machine may require replumbing. Changing the sequence of an EH system may require rewiring. In addition to the rewiring or replumbing is the time spent debugging and tuning the system with the new sequence. After a sequence is programmed into a PLC, it can be loaded in seconds and the debugging time is slight.

12.4.3 Remote Adjustment

In a hydraulically controlled system, remote adjustment of set points (from a control panel) may not be practical. It requires that the hydraulic valves and pump be mounted close to or on the control panel. In an EH or PLC system, remote set point adjustment is convenient with potentiometers or BCD thumbwheel switches.

12.4.4 Precision of the Set Point Adjustment

A precise adjustment of a pressure or flow set point may not be possible with a hydraulic control system. This is because the adjustment knobs of hydraulic valves are not graduated as digital potentiometers or BCD switches are. A precise adjustment of a hydraulic relief valve can be made, but this requires a pressure gage in the system.

12.4.5 Maintenance Personnel and the Selection of a Control System

To succeed in selling or applying systems, the service and maintenance of the system must be provided for. Computer control is a technology that is very sensitive to the knowledge level of maintenance personnel.

Hydraulic control has an advantage because of this in most plants. Maintenance personnel with machinery repair (mechanical) background are easily taught hydraulic system basics. Electrical principles are understood by fewer people, and computer skills by fewer yet. However, this is not the case in all plants. Industries that are electrical in nature often have many people available with electrical skills and computer skills.

The choice of a control system can be influenced by the work rules of the plant into which the system is going. A plant with strict work rules may have rigid rules regarding the job classifications of the workers and the work that they are permitted to do. An extreme example is a microcomputer-controlled system that requires workers of three job classifications to service it—a computer person, an electrician, and a mechanical person. In such an environment a hydraulic control system may be desirable because one person can do the entire job.

The point is that the background of the maintenance people can seriously impact on the success of a control system. Expecting people to learn new technology quickly is a very optimistic attitude. A system that cannot be maintained or repaired quickly is as useless as a poorly designed system.

12.5 IMPLEMENTATION OF A MICROCOMPUTER-CONTROLLED HYDRAULIC SYSTEM

Every designer has his or her own style and methods of designing hydraulic systems. Presented here are a few steps and tips that the

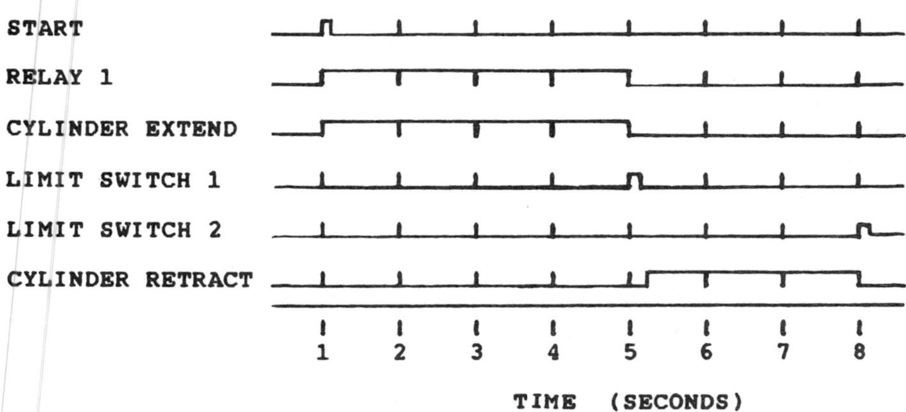

Figure 12.14 Example of a duty cycle diagram.

author has found useful in designing a hydraulic system controlled by a microcomputer.

12.5.1 Analysis of the Task to Be Done

The task to be done, or the process, must first be analyzed to size the actuators, the valves, and the pump. This consists of mainly hydraulic considerations, such as operating pressure, force on the cylindrical rods, hydraulic motor torque, and flow rates through all sections of the circuit. A preliminary hydraulic circuit diagram can then be done (Figures 12.1, 12.2, 12.3, and 12.7).

12.5.2 Defining the Duty Cycle

In this step a duty cycle diagram, as shown in Figure 12.14, can be drawn. Note which of the parts of the duty cycle are sequence dependent and time dependent. If it is important that an event occurs after a specific event, regardless of when it occurs, it is a good idea to program it this way.

12.5.3 Consideration of Various Control Methods

Hydraulic Control

Begin by considering hydraulic control without electrically operated valves. A simple system with a duty cycle that does not change is a good application for all-hydraulic control.

Sequence valves can be used for controlling the sequence of operation of the system. Pilot-operated valves are useful because the pilot pressure can be taken from various parts of the system to provide some hydraulic logic capability.

Electrohydraulic Control

Electrohydraulic control should then be considered. As systems become complicated it becomes difficult to control the entire duty cycle hydraulically.

A first step is to consider limit switches, as in Figure 12.2. These can simplify a hydraulic system by control of the sequence and set point control. Limit switches also provide timing capability to the system. This can be done by changing the cam that activates the limit switch, the location of the limit switch, or adding a time delay to the switch. (Limit switches and relays are available with built-in adjustable time delays.) If the timing of the sequence must be precise, photoelectric switches or proximity switches can be used instead of electromechanical limit switches.

Proportional electrohydraulic valves and pumps should be considered if multiple flow or pressure presets are required. The use of these devices may reduce the number of components in the system.

Microcomputer Control

This method should now be considered. The main advantage is programmability. If the duty cycle is complicated or it is changed often, microcomputer control is the obvious choice.

Designers should not limit themselves to the preliminary circuit design that they have already done or to the electrohydraulic circuits in the previous step. Computer control opens new possibilities for circuit design.

Timers can be programmed and activated by limit switches or events in the duty cycle. A more complicated duty cycle is made more practical by the microcomputer, so the designer may be able to design a more sophisticated process with higher performance than the preliminary, simpler systems that were considered.

Programmable set points and calculated set points are an important capability and should be considered.

12.5.4 Combinations of the Control Methods

The best control solution may be a combination of the three methods. There are tasks that may be better controlled electrically, and those that are better hydraulically controlled, within the same system.

It is usually better to control safety components and such functions as relief valves, emergency shutoff switches, and jog controls in the most simple (reliable) method available. Manually operated valves or mechanically adjusted valves are the most suitable.

12.5.5 Implementation

After the control method is chosen, the duty cycle diagram must be redrawn. At the same time the duty cycle is being drawn, the electrical and hydraulic circuit schematics must be done. It is important to do these at the same time because this helps the designer to decide if tasks should be done in hardware or software. It also saves the designer time if the three are done together, by eliminating the time it takes to get back into the details of the cycle.

12.5.6 Programming Tips

Regardless of the language used there are a few programming practices that can save time and effort in the implementation of a system.

Identify which functions are timing dependent and sequence dependent. Program these accordingly. If an event is sequence dependent (dependent upon an event occurring before it) and it is programmed based upon time, errors can occur in the operation of the system if an event does not occur on time. Events can occur at the wrong times owing to valve malfunction, leakage, or incorrect flow set points. If multiple events that are sequential are programmed based upon time, the sequence can be thrown off with unpredictable errors.

Flowcharts should be used prior to programming the code. These make debugging the program easier and help in documentation.

The program, regardless of the language used, should be structured to be easy to understand and modify. Avoid a seemingly clever code that appears compact but is difficult for another person to understand. Modularity is always a good idea. It will be necessary to change the program during initial system debugging and in the future, months or years after installation.

After programming, it is a good idea to test and debug the program as much as possible before operating the complete system. This saves time in troubleshooting to determine if a problem is software or hardware related.

12.5.7 System Start-up

If possible, it is a good idea to test and set the hydraulic components of the system prior to attempting to execute a complicated sequence. It gives the designer a chance to find faulty components in the beginning, simplifying debugging of the complete system. System troubleshooting can be complicated and overwhelming at times if one or more problems occur and interact.

Index